Praise for *From Farms to In...*

"I am very enthusiastic about holding up these stories to inspire others ... using their STEM backgrounds to make positive impacts in our communities and our world."
—**Karen Ross**, Secretary of the California Department of Food and Agriculture

"Amy Wu tells the stories of the underestimated women who are the driving force and changing face of agriculture in California. This book will serve as inspiration for the countless minority women who seek to be heard in rural America."
—**Christine Chavez**, granddaughter of Cesar Chavez and farm activist

"Finally, there's a book that celebrates the incredible women who will inspire the next generation of fabulous female foodpreneurs."
—**Gail Becker**, Founder and CEO, Caulipower

"Amy Wu is a masterful storyteller and voice for the unsung heroes of agriculture today."
—**Tonya Antle**, Cofounder and Executive Vice President of the Organic Produce Network

"Amy Wu has written a book that not only celebrates women innovators and entrepreneurs in food and agriculture, but also inspires the next generation of women to follow in their path."
—**Kara Goldin**, Founder and CEO of Hint, Inc., and author of *Undaunted: Overcoming Doubts and Doubters*

"I have seen young women from around the world coming up with creative solutions to food system challenges. Their approach is refreshing and urgently needed to get rid of polarized thinking. *From Farms to Incubators* injects inspiration and empowers the next generation of female leaders in the food and agriculture space."
—**Christine Gould,** Founder and CEO of Thought For Food

"By telling the stories of amazing women in agtech, like Amy Wu does in this book, we can help to prove that different approaches are to be celebrated and can yield great success."
—**Louisa Burwood-Taylor**, Head of Media and Research for AgFunder

"Amy Wu has done a phenomenal job collecting the stories of women founders in agtech. Their creativity, vision, and passion for using technologies to solve some of the biggest challenges that farmers face is something that we celebrate."
—**Carolyn Leighton**, Founder and CEO, Women in Technology International

"Amy's book of women in agriculture beautifully illustrates the diversity, passion, and commitment of an amazing group of women who are remaking modern agriculture through innovation and sustainable practices."
—**Mareese Keane**, Cofounder, Opengate

"*From Farms to Incubators* not only shines a fascinating spotlight on a growing and evolving industry but is another necessary catalyst for gender equality, providing a platform to the often voiceless and inspiration for the next generation of female leaders in this space."
—**Juliet Scott-Croxford**, CEO, Worth Media

"In her book *From Farms to Incubators*, Amy Wu has captured the stories of agtech's future pioneers. Using their life experiences to inspire others, these women are breaking down barriers while holding onto family values."
—**Karen Washington**, farmer and activist

A farmscape of lettuce in the Salinas Valley, California. Photo courtesy of Dexter Farm.

FROM FARMS TO INCUBATORS

Women Innovators Revolutionizing How Our Food Is Grown

Amy Wu

Fresno, California

From Farms to Incubators

Copyright © 2021 by Amy Wu.
All rights reserved.

Images in frontispiece collage courtesy of Trav Williams/Broken Banjo Photography

Published by Craven Street Books
An imprint of Linden Publishing
2006 South Mary Street, Fresno, California 93721
(559) 233-6633 / (800) 345-4447
CravenStreetBooks.com

Craven Street Books and Colophon are trademarks of
Linden Publishing, Inc.

ISBN 978-1-61035-575-9

135798642

Printed in the United States of America
on acid-free paper.

Library of Congress Cataloging-in-Publication Data on file.

Contents

For Dad.

"A smooth sea never made a skilled sailor."
—Franklin D. Roosevelt

Creating Fertile Soil for Untold Stories to Grow

Foreword by Lia Huber

Amy Wu had been an investigative reporter for twenty years at major media outlets when she got a call from an editor at the *Californian* asking if she'd be interested in moving to Salinas to cover agriculture.

It was 2016, and the agriculture technology (agtech) sector was burgeoning in the region. Agtech ticked several boxes on Amy's interest list: business, government, agriculture, and technology. She said yes and headed to the Central Coast from the Central Valley.

But there's a backstory here of a rising tide, brought on by the creative thinkers in Salinas and amplified by Development Counsellors International (DCI). It would become the fodder for Amy's future.

Photo courtesy of Lia Huber

The Spark That Put Salinas on the Agtech Map

Three years earlier, in 2012, economic hardship struck the Salinas Valley—known mainly as the "Salad Bowl of the world"—when Capital One, the region's main employer, closed its offices and left nearly 900 people out of work.

Salinas's city leaders had already been looking to shift the region's economic trajectory, and Capital One's departure made the case for doing so even more compelling.

Reeling from the loss of one of the city's largest employers, the City of Salinas launched its agtech innovation ecosystem initiative with the threefold aim to attract and support entrepreneurs and start-ups; seed and grow educational and training programs; and rebrand and establish the region as an agtech hub.

City leaders brought together a strategic coalition of stakeholders with vast expertise, such as John Hartnell of Silicon Valley–based consulting firm SVG Partners and Bruce Taylor of Taylor Farms. And they brought in DCI to spearhead rebranding and outreach.

At first, it was an uphill climb. No one had heard of the term *agtech*, and Salinas didn't have a big research university to give it gravitas. But DCI persevered, and the first big hit came in 2013 when it brought a reporter from the *Financial Times* to Salinas, resulting in the front-page article titled "Silicon Valley Links with Salinas Valley to Make Farming 'Smart'."

Just a couple of weeks later, the *San Francisco Chronicle* followed suit and soon the trickle of press became a torrent, with major feature stories in CNBC, *Wired*, the *Los Angeles Times*, and more. As the narrative about Salinas's and the region's credibility as an agtech leader grew to a national— even global—scale, there became more and more to write about.

So much so that when DCI pitched Forbes Media on the idea of holding an agtech summit in Salinas (Forbes had been considering the Midwest) and connected the team with those spearheading the agtech innovation ecosystem efforts in Salinas, Forbes said yes. And in July of 2015, the first Forbes AgTech Summit was held in downtown Salinas.

Which means Amy arrived in Salinas at the perfect time.

Explosive Growth

"I've always been interested in innovation and how it impacts society," says Amy. Over the next three years, Amy witnessed—and covered—plenty of it. Like the opening (and prolific growth) of the Western Growers Center for Innovation and Technology incubator. And the highly successful CSin3 program, which has paved the way for students who may have fallen through the cracks (40% of the program's students are female, 93% are minorities, and 74% are the first generation in their family to attend college) to earn a bachelor's degree in computer science in three years from California State University, Monterey Bay, and Hartnell College.

Another big boost to agtech start-ups was the launch of the THRIVE Accelerator Program, a highly competitive four-month program culminating in a business pitch at the Forbes AgTech Summit and up to $200,000 in seed capital.

The more involved Amy got in the agtech scene, the more she began to wonder how many agtech startup firms were led or launched by women. Being Chinese American, she recognized that the entrepreneurial road was exponentially more difficult for minority women in a new field spanning two arenas—agriculture and technology—that had traditionally been dominated by men. "I kept wanting to ask these women, 'What's motivating you to climb Everest?'"

So, in 2017 she began seeking out minority women entrepreneurs in agtech, hoping to learn about them and tell their stories.

The Incubation of *From Farms to Incubators*

Amy sought out grants to fund a documentary, *From Farms to Incubators: Telling the Stories of Minority Women Entrepreneurs in Agtech in the Salinas Valley and Beyond.* And for the next several months, she searched for minority women who had started companies in agtech and listened to their stories.

"A common thread I found among these women entrepreneurs was that they were intent on wanting to use their skills and education and professional background to create something that made an impact," says Amy. "They're truly on a mission to make a difference and solve problems in the food system."

Which is a good thing, because the food system is facing unprecedented challenges, like climate change, shortages in water and arable land, and supply chain disruptions due to global pandemics, all while being expected to feed approximately 9.8 billion people by 2050.

Innovative technology applied to agriculture can help farmers make smarter decisions and do more with less . . . less land, less water, less labor. And diversity drives innovation. "When you have a diverse group of people, decision making is more robust and you get better outcomes and more creativity because you have different perspectives," says Pam Marrone, CEO and founder of Marrone Bio Innovations.

Two of the founders Amy profiled were Diane Wu and Poornima Parameswaran of Trace Genomics, who won the THRIVE Accelerator competition in 2016. They created a simple soil test and added machine learning to unlock powerful insights into the billions of microbes that exist in just one teaspoon of soil.

Ask any of the subjects in Amy's documentary and they'll tell you that not only can starting an agtech company as a woman be daunting, it can also be lonely. Women control only 7% of US farmland and own only 14% of farms.

Moreover start-up founders with funding are 80% male and 90% white—despite the fact that, as First Round Capital reported, companies with a female founder performed 63% better than investments with all-male founding teams—and it can seem downright impossible.

As Anna Caballero, California State assemblywoman, says, "Women don't get treated well unless there's a critical mass." Amy is helping create that critical mass in agtech.

In 2018 her documentary took center stage at the Forbes AgTech Summit and has been screened around the country, including at SXSW and Techonomy. In 2020 she was named one of the "50 Women Changing the World" on *Worth* magazine's "Groundbreakers 2020 list." She has now published the book you are reading, *From Farms to Incubators,* and a companion exhibition that launched at the National Steinbeck Center in November 2020.

Amy came to cover the burgeoning agtech scene in Salinas that DCI had helped to grow. Now she's growing a movement that will pave the way for the next generation of women agtech leaders to find better solutions for agriculture, for business, and for our future.

Lia Huber is the author of Nourished: A Memoir of Food, Faith, and Enduring Love *and the founder of NOURISH Evolution. A seasoned food and travel writer for major publications, she consults on branding and messaging for companies with a cause. This foreword was first published by Development Counsellors International (DCI) in May 2020.*

Agtech and the Spirit of Innovation

Foreword by Dennis Donohue

From Farms to Incubators is an important book. It highlights the increasingly important role technology will play in the future of production agriculture, and it introduces us to some of tomorrow's new agricultural business leaders. The women who are featured in this book are building new companies, providing new solutions, and paving the way for a new generation of technically savvy women.

Amy Wu's background as a journalist has served her well. Her coverage of Salinas City Hall exposed her to the importance of agriculture to the Salinas Valley and its impact well beyond the borders of Monterey County. What's exciting is that she discov-

Photo courtesy of Western Growers Center for Innovation and Technology.

ered that the women she's met are not only providing needed solutions, they are also expanding traditional wealth creation possibilities beyond owning land and facilities. Those same personal economic opportunities suggest very real possibilities for a rural economic renaissance by creating companies that operate in the heart of key farming production areas.

I have been involved in the agtech movement since its inception in the Salinas Valley and have had the good fortune to work with a number of the women and companies profiled in this book. Agriculture is a demanding business, and Amy features a group of women who are meeting the challenges of the marketplace. These women innovators recognize it takes more than technology. From Diane Wu to Martha Montoya to Pam Marrone, all of the women featured here understand they have to meet the business needs of their potential new customers. They have to solve problems, reduce costs, or create opportunities.

The agtech sector will not continue to grow without industry adoption. One of the traits that characterizes many growers is their ability to solve problems. These women share that trait, which will help agtech solutions come on line more rapidly.

Problem solving is critical to building a strong organization and, most importantly, being laser focused on the customer. Another key to success is recognizing those good ideas. Good teams all need a strong, informed relationship with their customer. One of the real keys to success is the ability to "cocreate" solutions between innovators and growers.

The role of the Western Growers Center for Innovation and Technology is to help accelerate new solutions for the wide array of challenges that currently confront production agriculture. I have directly worked with a number of the women Amy writes about and have observed firsthand their focus on getting to the heart of a problem, opportunity, or objective. It is a critical trait to be intent on being as specific as possible. It's how real progress takes place.

Beyond intellectual capacity, it requires patience and a willingness to manage the expectations of all the stakeholders involved in moving technology or innovation forward. I have always been struck by those women whom the center has had a chance to work with in various capacities.

The women featured in *From Farms to Incubators* demonstrate the diversity and excitement of agricultural opportunities for the next generation. These women have a passion for what they do, and that is the real key to ensuring American agriculture's future success.

Dennis Donohue has over thirty years of experience in the produce industry and has worked with a number of industry leaders. Donohue's strong civic commitment includes his leadership as Salinas mayor from 2006 to 2012. He is widely credited for his efforts to link the Salinas Valley with Silicon Valley while mayor. He has also served as the chair of the Central California Grower Shipper Association and as chair of the Salinas Valley Chamber of Commerce.

Introduction

"The woman who follows the crowd will usually go no further than the crowd. The woman who walks alone is likely to find herself in places no one has ever been before."

–Albert Einstein

Both agriculture and technology have a long history in California but until recently have rarely intersected.

Technology is focused on disruption and innovation, while much of agriculture is still managed with clipboards and pencils. A commonality is that women and individuals with diverse cultural and ethnic backgrounds have historically been underrepresented in both industries. Now, a new generation of start-ups led by women of diverse backgrounds is seeking to use innovative technology to provide novel perspectives and solutions to agriculture's problems.

From Farms to Incubators is a collection of visual and written portraits of women leaders and innovators in the growing sector that combines agriculture and innovation known as agtech. Most of the women profiled in this book are based in or have professional connections with California, which leads the United States in agricultural production. The state's ag sector is a $50 billion industry that produces over four hundred commodities.[1]

This book documents the stories of these women in their own words and through photos provided by the women and a select group of photographers in California who lent their talent to this special project. In the chapters to come the women share their unique journeys, including their challenges and successes.

Amy Wu. Photo courtesy of Bill Winters.

This leads me to my journey.

I am a first-generation Chinese American and consider myself transcontinental and bicoastal. I grew up and went to school in New York; I was a history major at New York University. I have family in Hong Kong, where I lived and worked in the 1990s and 2000s. Much of my extended family live in California, where I was based and worked for a number of years. I've been a writer and journalist for twenty-five years and counting.

My passion to unearth underreported stories and amplify new voices is what led me to highlight the stories of emerging communities in agriculture. In recent years my focus has been on telling the stories of women leaders and innovators in agtech. This includes all technology invented to help farmers be more efficient in their work and to increase yield and cut costs. I became interested in this topic when I was a newspaper reporter in Salinas, beginning in 2016, and my beat was agriculture and local government. In my wildest dreams I could not imagine that my journey to Salinas would lead me to making a film and writing a book about women in agtech.

The Salinas Valley serves as the backdrop for much of the book and offers a unique lens for sharing the stories of these amazing women. In the Salinas Valley, agriculture is a $9 billion industry, and over 80% of the country's leafy greens are grown in the region (hence the City of Salinas's apt moniker the salad bowl of the world.

By the time I arrived in early 2016, Salinas had already spent several years focused on boosting the agtech sector. The Western Growers Center for Innovation and Technology, a business space and accelerator for agtech start-ups, had launched in 2015, and the city hosted the Forbes AgTech Summit a year later.

At the same time, as a woman and a person of color, I noticed there were not many women, especially women who looked like me, at agriculture-centric meetings and events. I started to ask, "How many minority women entrepreneurs are there in agtech here? Do you know of any minority women entrepreneurs in agtech?" I was met with a lot of silence and nervous laughs. At one conference in Silicon Valley someone responded to my question with "You're looking for a unicorn, aren't you." They were referring to a business term pointing to a start-up valued at over $1 billion, a rarity just like the mythical animal.

I didn't know if there were any women who fit the bill, in fact, but I responded to a callout for stories about minority women business owners by the International Center for Journalists (ICFJ). I was awarded a grant in 2016, and slowly I uncovered one story after another of women innovators in agtech. That grant fueled a series of profiles that were first published in the *Salinas Californian* and in the award-winning documentary *From Farms to Incubators: Telling the Stories of Minority Women Entrepreneurs in Agtech in the Salinas Valley and Beyond*. A second grant in 2017 from the International Media Women's Foundation helped expand that documentary.

The documentary has been screened at dozens of venues, including SXSW, Techonomy, and EcoFarm, and served as a critical vehicle in launching a discussion about the role of women, especially women of color, in this next chapter of agriculture. I've now led and organized dozens of screenings and panels where we watch the film and hold a discussion with women entrepreneurs in the agtech space.

I became increasingly fascinated with the stories of minorities and women who are transforming the agriculture scene with technology. Agriculture, as I'd learned after all, is much more than tractors and overalls. It is a world of tremendous opportunity that includes science, research, marketing, and innovation. Going forward, it will require a workforce adept with a new set of skills and a knowledge-based economy.

I have discovered dozens upon dozens of women innovators in agtech, not only in California but globally. They are tackling some of the biggest challenges that farmers face: severe labor shortages, limited water and land supplies, and the pressure to feed what is forecasted to be close to 10 billion people globally in 2050.[2]

Four years later, *From Farms to Incubators* has grown from a series of stories into an initiative, a combination of storytelling and advocacy. The journey has taught me that there is great power in telling peoples' stories. By sharing and highlighting those stories, we give them a voice, we uplift new voices, and we fill in the gaps of what will someday be history.

From Farms to Incubators: Women Innovators Revolutionizing How Our Food Is Grown uses the power of storytelling to increase awareness and also hopefully ignite an engaging discussion over the various factors that are causing shifts in our food systems. Ultimately, I hope the book serves

A single sunflower (the first to bloom) amid a landscape of its peers. Photo courtesy of Amy Wu.

as a vehicle for inspiring youth—especially those from rural or underserved communities—to consider that agriculture extends far beyond tractors and overalls. It is a sector that offers an amazing range of opportunities and involves exciting and innovative technologies in the fields of software, hardware, drones, artificial intelligence, sensors, data analytics, blockchain, and robotics.

By sharing the stories of women leaders and innovators in agtech, especially women of color, I hope to ignite the "can-do" spirit in more women, especially younger women, so that they will launch their own enterprises in agtech, agbio, and foodtech.

As Miku Jha, founder and CEO in AgShift, says in the *From Farms to Incubators* documentary, "It has to start early on. My daughter is ten years old and she's already sold into agriculture, and she's extremely fascinated. I took her to certain farms and she has been seeing what I want to do. She asks me what is it that I'm doing, and when I share that with her, she has that spark in her eyes."

Although the chapters that follow spotlight innovation and how technologies are helping growers, it is the individual stories of the women, their personal journeys, their struggles, and why and how they continue to pursue their dreams that make up the heart of this book.

This book is a snapshot of a moment in time. Fifty years from now a young woman may pick it up, read the portraits and look at the profiles, and marvel at what the landscape was like. Hopefully, they and their daughters and generations after them will celebrate these stories as a part of the long history of agriculture and technology and remember the women who were pioneers in their own right.

—Amy Wu, August 2020

The information for this book was gathered from 2016 through 2020. For up-to-date news on the people profiled herein, please visit www.farmstoincubators.com.

Notes

1. California Department of Food and Agriculture, "California Agricultural Production Statistics," 2019, https://www.cdfa.ca.gov/statistics.

2. United Nations, Department of Economic and Social Affairs, "World Population Projected to Reach 9.8 Billion in 2050, and 11.2 Billion in 2100," June 21, 2017, https://www.un.org/development/desa/en/news/population/world-population-prospects-2017.

INNOVATION MEETS AGRICULTURE

The landscape between two worlds overlaps on US Highway 101, the major thoroughfare connecting San Jose and Salinas in Northern California. Modern campuses of tech giants and office spaces blur into municipalities famous for their crops—Gilroy for garlic, Watsonville for strawberries—and then the two worlds meld into a vast farmscape of lettuce.

Enter Salinas, the county seat for Monterey County and the birthplace—and novelistic backdrop—of renowned novelist John Steinbeck. In Salinas, which bears the moniker "Salad bowl of the world," agriculture is front and center. The region's $9 billion ag industry produces approximately 80% of leafy greens in the US.[1]

Fieldworkers picking and packing lettuce at a farm in Salinas. Photo by Dexter Farm.

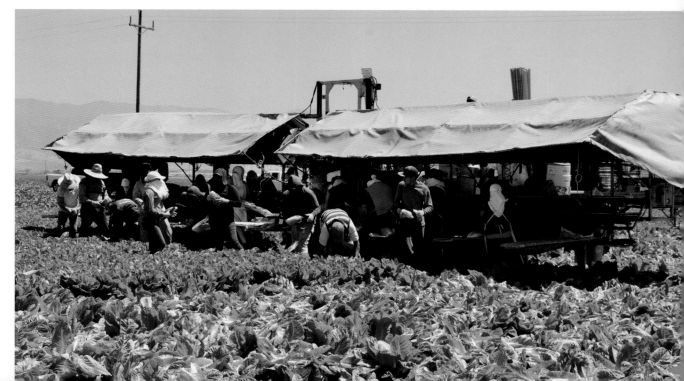

"The future belongs to those who believe in the beauty of their dreams."
— Eleanor Roosevelt

The local economy of the region—which is also made up of the cities of King City, Greenfield, Soledad, and Gonzales—is tied strongly to agriculture.

It also serves as headquarters to some of the country's largest agriculture companies and producers—including Taylor Farms, Earthbound Farm, Fresh Express, and Driscoll's, for example—that supply America's restaurants, grocery shops, and retailers.

Salinas is also a mecca for the fast-growing agriculture technology (agtech) sector, which marries agriculture and technology. The vast spectrum of technologies found in agtech comprises [or includes] artificial intelligence (AI), widely defined as the ability of a computer or a robot controlled by a computer to do tasks that are usually done by humans; blockchain, a growing list of records, called blocks, that are linked using cryptography; and everything from "big data" to robotics, automation, drones, and mobile apps.

Agtech refers to any innovation that is specifically created to tackle the array of challenges that growers are wrestling with, such as climate change, limited water and land supply, a severe labor shortage caused by fieldworkers aging out, and a depletion of healthy, farmable soil. At the same time there's a dearth of young people who are entering the field of agriculture, in part because land overall is expensive as is farming itself. The sector always grapples with a number of

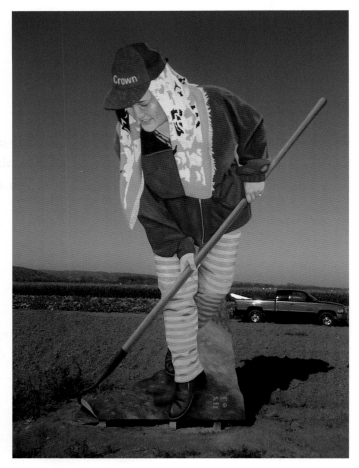

An art piece depicting a farmworker. The artwork, painted on painted plywood, was created by John Cerney, a renowned Monterey County-based artist. Photo courtesy of Dexter Farm.

The Western Growers Center for Innovation and Technology, based in downtown Salinas, is a business center for agtech start-ups. Photo courtesy of WGCIT.

uncertainties, many of which come down to the caprices of Mother Nature. And on the macro level, growers face pressures to produce enough food for a global population that is forecasted to hit 9.8 billion in 2050. Agtech seeks to help farmers deal with all of these issues and to make agriculture more profitable and efficient at the same time.

Since 2015 Salinas has been the home of the Western Growers Center for Innovation and Technology (WGCIT), an incubator for agtech start-ups that is located in the city's downtown. You'll find the place buzzing with entrepreneurs meeting with agriculture companies or connecting with investors. A new vibrancy defines downtown too, with an array of cafés, chic restaurants, and boutiques. Main Street is anchored.the modern Taylor Farms Building, reflecting the growth of the local economy. It is the unique connection and geographical proximity to ag and innovation that make the Salinas Valley an attractive (if not obvious) ground zero for agtech.

Aubrey Donnellan, CEO of Bear Flag Robotics, is one of the entrepreneurs you might encounter in downtown Salinas. Her company, which makes autonomous self-driving tractors, joined the WGCIT in 2018 to be closer to the growers.

"It's like a fast track to be connected to growers who have a vested interest in technology and are implementing their technology on their farms," says Donnellan soon after her company joined WGCIT. "It streamlines the relationship-building process. In the Bay Area we don't have natural opportunities to network with the customers." Bear Flag Robotics is just one of roughly sixty agtech start-ups that were based in the WGCIT at the start of 2020, a significant uptick from the handful present at the 2015 launch.

From left, the late Joe Gunter, former mayor of Salinas; Ray Corpuz Jr., former Salinas city manager; and Bruce Taylor, the chairman and CEO of Taylor Farms, at the Forbes AgTech Summit in 2015. Photo courtesy of the City of Salinas.

Serendipity

This book explores why Salinas is a case study on the critical significance of innovation and technology in food and farming as we continue to move forward in the twenty-first century. This book also highlights the stories of the entrepreneurs who have devoted their professions and lives to helping solve some of the biggest challenges that growers face. In order to understand the personal stories of women leaders in agtech, we must consider how the agtech sector developed in Salinas and the importance it holds for the region.

Salinas's agtech ecosystem is a collaborative effort among the City of Salinas, SVG Partners (the Silicon Valley–based consulting firm the city tapped to grow the sector), and the WGCIT, the powerhouse of the agriculture industry and the pipeline of agtech start-ups. Together these entities make up the engine that has fueled this fledgling sector led by a new generation of entrepreneurs who reflect the diversity of the changing workforce across all industries in the United States.

Although the creation and development of the Salinas agtech ecosystem appears to be part serendipity, it also isn't accidental. Turn back to 2012, when the city lost Capital One, its major employer, because of a company acquisition. Nearly nine hundred people were out of jobs. City leadership, including then mayor Dennis Donohue and newly appointed city manager Ray Corpuz Jr. made a decision to focus investment and energy on building the agtech sector. It was then that the city hired and later retained SVG Partners to develop the city's agtech sector.

SVG's CEO and founder John Hartnett says the original strategy, scope, and framework were called the Steinbeck Innovation and was based on four pillars of development: education (training and advanced research), start-up acceleration, investment, and corporate strategic engagement.

Education involved programs such as CoderDojo Hartnell College, which teaches computer coding to youth. Hartnett adopted this educational program from a similar one held in Ireland. The program was established in 2013 by CoderDojo Foundation by cofounder James Whelton, who envisioned children worldwide having the opportunity to learn code and to be creative with technology in a safe and social environment.

SVG also conceived of holding a local agtech summit and partnered with Salinas and several big agriculture companies such as Taylor Farms to hold one in the summer of 2014. Hartnett and Salinas eventually pitched the idea to Forbes Media, beginning a relationship that continues to this day. The Forbes AgTech Summit is held annually in Salinas and is one of the premier gatherings for leaders, executives, investors, and entrepreneurs in the agritech sector with a goal of connecting them.

Corpuz said the current agtech ecosystem is "much more robust and more than the four pillars." Since Salinas's initiative to launch the agtech sector began, many more companies and academic institutions have jumped on the agtech bandwagon.

The city of Salinas is home to major agriculture companies, including Taylor Farms, Fresh Express, and The Nunes Company Inc. (owner of the Foxy label). Photo courtesy of Dexter Farm.

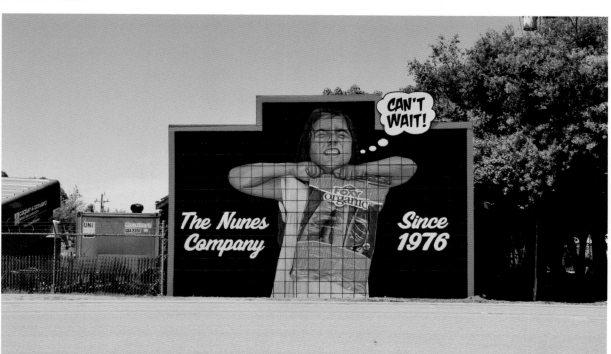

The influence of agtech has helped people involved in the local economy steadily shift away from a purely traditional way of doing things to new ways of thinking.

Steve McShane, a small-businessman who owns an organic fertilizer facility, is also an entrepreneur who launched a start-up that sells insect fertilizer. He has served on the Salinas City Council since 2010 and has been a staunch supporter of the agtech ecosystem.

"We've got mounting challenges," says McShane, noting the severe labor supply shortage and ongoing water and land supply management issues. Since 2012, and especially in recent years, the agriculture industry—known to be skeptical about outsiders and slow to adopt new technologies—is increasingly receptive to and keen on new solutions.

"The naysayers have really turned," McShane said, referring to growers. "I'd say that local culture has broken through those traditional barriers pretty good. The dance is definitely taking place in both directions," he said of growers, innovators, and entrepreneurs.

Salinas is also known for its annual rodeo held at the Salinas Sports Complex. A sign welcomes visitors to the city. Photo courtesy of Dexter Farm.

Donohue, who has been leading the WGCIT since 2016, said the innovation center has served as a talent pool for growers seeking new technologies. The start-ups reflect a wide range of innovation from robotics and automation to data mapping and soil testing. The WGCIT also serves as a resource and often a base for finalists of the THRIVE Accelerator program, which is organized by SVG Partners. The program chooses a cohort of agtech start-ups annually, providing funding and resources, and connects the company founders with growers and large agriculture companies in the Salinas Valley.

The Salinas Valley is the leading producer of leafy greens in the United States. Photo courtesy of Dexter Farm.

Tinia Pina, founder and CEO of Re-Nuble, a Brooklyn-based company that supplies chemical-free hydroponic fertilizer, was one of the THRIVE Accelerator participants in 2018. Pina says she applied to be more closely connected with the agriculture community outside New York City.

"I have found the investment community on the West Coast to be more tolerant of the scaling timeline typically involved in agriculture-related ventures," Pina says.

For Salinas, the burgeoning agtech sector means responding to the needs of these new businesses. In April 2019 the city passed a policy that gives itself permission to develop and upgrade its wireless network. Andy Myrick, the city's economic development director, said this policy means that the city is asking itself forward-looking questions such as "Do we have the fiber we need in place? Are we ready for 5G? Do we have physical spaces for these companies?"

In June of that year, Main Street of Salinas was abuzz with the Forbes AgTech Summit, which attracted nearly one thousand executives, investors, and entrepreneurs in the agtech space, many of them from Fortune 500 companies. At the summit, Donohue—himself a seasoned veteran in the ag industry—reiterated the point that developing the agtech sector is a necessity for economic vitality.

The agriculture and technology industries will increasingly compete for talent as both require workers with a new set of skills. "The future economy and future high-skilled jobs are going to include a lot more tech," Donohue says. "I think human capital will be a value proposition."

The pipeline of talent that Salinas aims to cultivate to fuel agtech continues to move forward. In 2019 Salinas launched a new workforce program for agtech job training in partnership with Hartnell

College in Salinas and the college's satellite campuses in neighboring cities such as Gonzales. And that same year, Digita NEST, a nonprofit headquartered in Watsonville that trains youth ages fourteen to twenty-four in tech skills such as programming, game design, and graphic arts, opened a satellite office in Salinas with the goal of preparing youth for the growing number of agtech jobs right in their home region of the Salinas Valley. Since the workforce program was launched, over 2,500 youth have gone to Digital NEST for services training.

"Developing and supporting the development of the agtech ecosystem in Salinas is our number one economic strategy. We are committed for the long term. We believe the benefits will be more investments and jobs for Salinas," city manager Corpuz said.

What Has Emerged

Salinas is a microcosm of what is happening across the US and throughout the world. Growth in both the agtech and foodtech sectors is picking up pace: in 2017 the industry saw investment rise with an uptick in investments and number of start-ups. According to PitchBook, 2017 saw $1.5 billion in agtech investment, with over three hundred investors and over 160 deals, compared to 31 deals and less than $200 million invested in the sector in 2007.

Two years later, in 2019, foodtech and agtech start-ups raised $19.8 billion in venture funding across 1,858 deals, according to AgFunder, a leading news source for the agtech sector. The start-up

Lettuce fields in Salinas. Photo courtesy of Dexter Farm.

The Taylor Farms Building, headquarters of Taylor Farms and the Western Growers Center for Innovation and Technology, anchors downtown Salinas. Photo courtesy of Dexter Farm.

investment in this space has grown 370% since 2013, reaching $4.7 billion for 695 deals in 2019, according to AgFunder's 2020 Farm Tech Investment Report.[2]

Out of this, something remarkable is emerging: opportunity for women innovators and entre-preneurs. Up until now Agtech has been dominated by white men, but a number of the agtech start-ups—including several inside the WGCIT—are being launched and led by women, especially those of color. While their stories are diverse, they share numerous commonalities. The majority enter the field with a strong STEM (science, technology, engineering, and math) background, and they share a passion to create technology and products that can improve food and farming systems.

Their drive is also ignited by challenges and opportunities. Farmers around the world wrestle with everything from water and land supply shortages to labor shortages and keeping family farms running as younger generations consider other professions. The majority of family farms have found it hard to stay open. The reality is that the average age of farmers continues to rise, according to the latest census from the US Department of Agriculture. This leaves open the question, Who will inherit the land and farming as livelihood?

A Cultural, Societal Shift

One group that has helped answer that question is a new generation of women, including women of color, with a passion for farming who have launched or are leading start-ups that are seeking to solve agriculture's problems with tech innovation.

The reality is that a need for talent coupled with the changing demographics of the twenty first century is morphing the workforce into one that is diverse in gender, race, and ethnicity. The large scale of agriculture in the Salinas Valley creates a tremendous opportunity for the next generation. Agtech opens doors of opportunities for developing a twenty-first-century homegrown knowledge-based workforce that has the skills to tackle these challenges. And young women who have an interest in STEM and in food and agriculture may be much in demand.

Lorri Koster, former CEO and chairwoman of the Salinas-based agriculture company Mann Packing, noted during a discussion with me that it is critical to convey to the next generation of women that agriculture is much more than "tractors and overalls."

Back in 2018 in an interview with Glenda Humiston, University of California vice president of Agriculture and Natural Resources, at the organization's headquarters in downtown Davis, we talked about her passion for promoting agtech, building a skills-based workforce, and fostering a new generation of growers and entrepreneurs. Humiston noted that "part of the problem is when you talk about agriculture, people think there are only two jobs there, farmer or farmworker. In reality, there are thousands of jobs in the ag ecosystem—laboratory people, irrigation specialists, mechanics, people operating drones, and more. We've got to get people to understand that there are some very cool, very fun, and pretty well-paying jobs out there in that sector."

Humiston sees a tremendous opportunity in the sector for women passionate about agriculture, technology, and food. Women care about health, wellness, and nutrition, and there is a cultural shift too, as more grower families are as likely to hand the land over to daughters (who are interested in farming) as to sons. As an example, Salinas Valley hosts a successful program called CSin3, a collaborative degree program between Cal State Monterey Bay and Hartnell College. It builds a pipeline of new talent with a focus on young women and those from underserved communities. The CSin3 program offers a computer science degree in three years, and boasts a 70% graduation rate where 40% of the students are female and 93% are an underrepresented minority. One of the program's successful alumni is Rivka Garcia, who worked at the Salinas–based agtech company HeavyConnect before moving to Mann Packing as a business analyst/developer.

Humiston called positive changes such as these an "important societal, cultural shift" and is leading efforts to improve collaboration throughout the University of California, California State University, and California Community College systems to expand such programs. And on the municipal side, Corpuz said he hopes the agtech ecosystem will be extended to focus on diversity

A vast field of lettuce spreads out into the Salinas Valley. Photo courtesy of Dexter Farm.

and inclusiveness, such as recruiting more women entrepreneurs. "The issue of equity gender or race is very big in the whole agtech space. We need to be mindful of that," he said.

The Search Begins

When, where, and how did the journey to write about women founders in agtech start?

In autumn of 2016, I began searching for minority women entrepreneurs in agtech in Salinas when I was reporting for the *Salinas Californian* and assigned to cover agriculture and government.

I asked, "Do you know of any minority women entrepreneurs in agtech?" at various events and settings, including the following:

- In early 2017 at the THRIVE AgTech Summit—an industry-focused event held in Menlo Park, California—a videographer and I visually scouted out potential subjects. The first woman I encountered was Erica Riel-Carden, at the time an agtech attorney at the Royse Law Legal Incubator and now an investment banker at Global Capital Markets in San Francisco.
- At the WGCIT I met with Dennis Donohue, who pointed me to Diane Wu (no relation to me) and Poornima Parameswaran, cofounders of Trace Genomics, a soil-testing company

that opened a Salinas office in 2018 that had produced a soil-testing kit for analyzing soil DNA.

- I also met with Pam Marrone, founder and then CEO of Marrone Bio Innovations, a NASDAQ-listed company that produces bio-based pesticides. She was mentoring women entrepreneurs in agtech. Marrone connected me with Miku Jha, founder of AgShift, which at the time was developing a software similar to Quickbooks (accounting program) for growers. Marrone also referred me to Thuy-Le Vuong, founder of The Redmelon Company. Vuong had invented a technology that extracts the oil from gac, an exotic fruit high in beta-carotene. Marrone then linked me with Fatma Kaplan, CEO and founder of agbiotech start-up Pheronym, which uses pheromones for eco-friendly pest management solutions, including controlling parasitic roundworms called nematodes.

As you can see, in unearthing one source, I found another and then another, and as I write this I am finding more women involved in agtech in California, across the country, and around the world.

In mid-2017 after I released an initial documentary, *From Farms to Incubatators: Telling the Stories of Minority Women Entrepreneurs in Agtech in the Salinas Valley and Beyond,* spotlighting a select number of women's stories, and held screenings and panel discussions, the most common reaction I received from the subjects I interviewed was one of awe. "I didn't know they existed until now," the women would tell me, requesting contact information for follow-up. Nearly three years later, curious to know where they were now personally and professionally, I circled back with the women whom I first interviewed. This is a sampling of what I found:

- Diane Wu and Poornima Parameswaran, cofounders of Trace Genomics, had raised $13 million in a series A round.
- Christine Su, founder and CEO of PastureMap, which offers mobile mapping apps to cattle farmers, had expanded staff and moved their base into the Impact Hub SF building. Su is a sought-after speaker at industry conferences and chaired the Grassfed Exchange Conference in April 2019 at the Sonoma Fairgrounds.
- In early 2019, Miku Jha, founder and CEO of AgShift, unveiled the company's latest product, an autonomous food inspection system machine for produce and nuts.
- Jessica Gonzalez, a cofounder of the Salinas-based agtech start-up HeavyConnect, moved back to her native Merced to join her nine siblings to care for their ailing father and the family farm. Culture and family played a pivotal role in the decision, but Gonzalez is far from giving up on her innovative dreams. She used her HeavyConnect experience to introduce mobile payment on the family farm for employees and to keep track of expenses. A budding entrepreneur, she also launched Happy Organics, which sells cannabidiol-infused honey produced by the bees bred on her family farm.

During my research for this book, I connected with many other female agtech entrepreneurs including ninth-generation farmer Penelope Nagel, COO of San Diego–based Persistence Data Mining, and Mariana Vasconcelos, CEO and founder of Agrosmart, a Brazilian company that creates software that employs artificial intelligence to better predict factors such as soil quality for farmers.

In early 2020, I reflected on what is now a three-year journey and noticed the uptick of women leaders in agtech. One is Ponsi Trivivaset, CEO of Inari, a Boston–based company that is creating a customized seed that adapts to geography, altitude, climate, water supply, and soil quality. Another is Murielle Thinard-McLane, who with Andrea Chow coheads Ontera, a Santa Cruz–based company with a diagnostic platform that examines DNA and RNA in plants, which targets growers.

Not limited to the Salinas Valley, I had the pleasure of meeting Ellie Symes, CEO and founder of The Bee Corp. Symes is a twenty-something rising star in the agtech world. In March 2019 The Bee Corp. was chosen to be a member of the THRIVE Accelerator program.

The Western Growers Center for Innovation and Technology, also known as WGCIT, houses over fifty agtech companies, many of them start-ups. Photo courtesy of WGCIT.

A Delicate Balance

Despite the rapid advances in science and technology, the impact of agtech on agriculture is still heavily affected by Mother Nature. The realities of climate change, dramatic temperature fluctuations, and the constant struggle of balancing agriculture with conservation and wildlife ecology can mean a year or more of crop failure. California continues to face a land and water shortage and shockingly sudden climatic changes, as 2020's wildfires underscore. Many growers wrestle with the rising cost of doing business in an industry that's still highly dependent on labor and machinery.

Al Courchesne, a farmer who founded and owns Frog Hollow Farm, an organic farm in the town of Brentwood, just east of San Francisco, changed his crop plan because of shifting weather patterns. Declining chill winter hours, prompted by persistently warmer weather, resulted in stunted fruit blossoms. This compelled Courchesne to graft the existing trees with new varieties that require fewer chilling hours. Extreme storms triggered fungal diseases that destroyed his apricots and other stone fruits. In recent years, he stepped up the dried-fruit business and spent money to tackle the labor shortage. "We invested over $100,000 in equipment to get the peels off the peaches and the stems off cherries," Courchesne says.

Courchesne launched his first orchard in 1976 and began farming with organic practices in 1989 as part of a commitment to build soil health and community. He is keen on data and analysis about soil that is essential to crop yield. As a result, he is open to conversations with technology companies, the majority of them start-ups that have approached him with their products. An imaging drone, for example, might serve as "eyes on the ground" and assist Frog Hollow's pest control advisor. "It would empower us," says Courchesne, who needs help since the farm has grown from 140 to 280 acres. "The drone technology is really good—we're looking to adopt that as soon as possible, and somebody told me that drones will be picking peaches in the near future." He's keen on "any new technologies that would help us understand soil better." He has heard from a number of companies offering farm record data software.

Donohue says the growing number of start-ups at the WGCIT reflects the need of growers to turn to technology. Some innovations are particularly needed: automation for planting, harvesting, thinning and weeding; technologies that offer data and analysis for food safety and soil health; and innovation catering to specialty crops—like celery, berries, and pistachios—that have a higher level of complexity. The sky is the limit for those with knowledge, creativity, passion, a solid business plan, and ,importantly, proven technology.

Changing Times

A paradigm shift, a fundamental change in approach or underlying assumptions, can take generations to take hold, but the women profiled in this book are steadily making a mark with their technologies. It is my hope that sharing their stories will ignite the "can-do" spirit in more women—especially younger women and women of color—and inspire them to launch their own enterprises in agtech. No doubt the innovations they are creating will continue to make a difference and spark a trickle-down impact from seed to table. Food is a necessity and deeply intertwined in our lives. Food, for many of us, connects us with loved ones and community and to our cultural heritage.

The following portraits tell a story about innovation, entrepreneurship, and the tenacity and a new generation of women who are seeking to make a mark in food and farming systems. Each story is also ever evolving and affected by current events, geopolitical factors, and consumer sentiment. As with any emerging sector, there is trial and error, and as with the start-up world, many companies fall by the wayside and a few will survive and thrive. The creation and refinement of agtech innovation

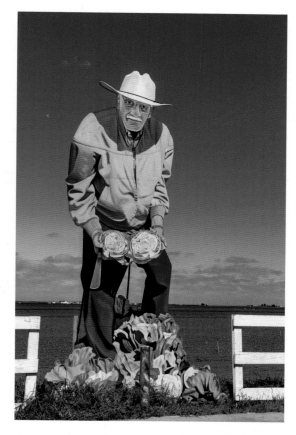

An art piece depicting a farmer. The artwork, painted on painted plywood, was created by John Cerney, a renowned artist based in Monterey Bay. Photo courtesy of Dexter Farm.

involves connecting with growers, working closely with consumers, building a team, and raising capital. Will the ripples eventually lead to a paradigm shift in the sector and also in who leads it? The stories of women agtech innovators in the Salinas Valley and beyond suggest the answer may be yes.

DEVELOPING AN AGTECH ECOSYSTEM

–contributed by Ray Corpuz Jr., former city manager, Salinas

In 2012 the City of Salinas initiated plans to transform the salad bowl of the world to the "agtech hub center of excellence." In developing a successful agtech ecosystem, the city pursued and developed seven major key elements..

They included corporate commitment, city sponsorship, entrepreneurs, university research, capital investment, networks, and marketing. During formation, the city did not include diversity and inclusion as key elements of the agtech ecosystem. The active and open innovation platform used to drive participation and commitment did not visibly promote and support diversity and inclusion. We have learned from that early start a valuable lesson.

A healthy agtech ecosystem comprises a diversity of people and pro-

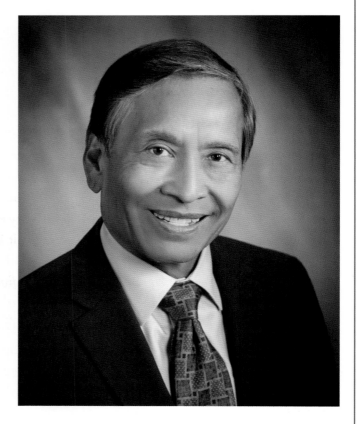

Ray Corpuz Jr. Photo courtesy of the City of Salinas.

vides economic opportunity for entrepreneurs and workers in the system. Entrepreneurs and workers can come from around the region and from other areas of the country. They can also come from other countries. What matters is that leaders in the agtech ecosystem assess, measure, and develop strategies to ensure that diversity and inclusion are visible and demonstrable throughout the ecosystem. We know that US demographics are shifting and that by 2040, 13 states, 103 metros areas, and 602 counties will have majority nonwhite populations.

Innovation is a collaborative process and relies on a diverse set of players defined by race, gender, and other characteristics to create new ideas and products. There is a positive relationship between diversity and innovation in business systems.

The intersection of agriculture and technology is one of the fastest-growing sectors in the Salinas Valley. Salinas is an agtech epicenter where innovation is being developed and produced from AI [artificial intelligence] to robotics. Agtech is also one of the fastest-growing sectors in Silicon Valley. In 2019 it totaled over $19 billion. However, women entrepreneurs received a small share of these investments. And women-led agtech companies represented only 2% of venture capital investment that same year.

Given the vast populations affected, we need to be sure women and minorities are engaged in the agtech ecosystem. In Salinas, 77% of the residents identify as Hispanic. Women make up 51% of residents. Beyond the economic considerations we have a moral obligation to increase the number of women and minorities in the agtech space. We are starting to see visible evidence of diversity and inclusion in programs. However, leaders in the agtech ecosystem from the corporate world and the venture capital financiers, as well as entrepreneurs and government, must build a visible and firm commitment to diversity and inclusion for women and minorities. It is imperative for progress. It is a key element for the continued development of a successful agtech ecosystem. Most importantly, it is the right thing to do.

Ray Corpuz Jr. served as the city manager for Salinas from 2012 through July 2020. Corpuz has been in local government for over forty-eight years, including serving twenty-nine years as city manager for the cities of Seaside, California, and Tacoma, Washington, in addition to Salinas.

Citations

Richard Smoley, "Salinas: Innovation by Necessity," Blue Book Services, March 10, 2020, https://www.producebluebook.com/2020/03/10/salinas-innovation-by-necessity.

Amy Wu, "Innovative Women in Agtech Play Pivotal Role in Salinas Valley," December 19, 2018, https://agtechsalinasca.com/2018/12/19/innovative-women-in-agtech-play-pivotal-role-in-salinas-valley.

Notes

1. David Castellon and Amy Wu, "Monterey County Ag Reports Record Year," *Salinas Californian*, June 28, 2016, https://www.thecalifornian.com/story/news/local/2016/06/28/monterey-county-ag-reports-record-year/86485684.

2. AgFunder, *2020 Farm Tech Investment Report*, https://agfunder.com/research/2020-farm-tech-investment-report.

Woman, by JC Gonzalez, artist, farmer, and educator in Salinas, California. Mixed-media on canvas with acrylics and oils. "This piece is a mixed-media collage. I included butterflies as a symbol of transformation but also for all the woman immigrants in this country. I also integrated photos from the women who were featured in the National Steinbeck Center exhibition."

POORNIMA PARAMESWARAN AND DIANE WU

Trace Genomics, Burlingame, California

Two Stanford PhDs Found Success in Creating a Soil-Testing Kit for Growers

From left, Trace Genomics CEO Dan Vradenburg, Diane Wu, and Poornima Parameswaran. Photo courtesy of Trace Genomics.

A potluck gathering of members of Andrew Fire's lab and their significant others. Fire did not attend and is not pictured, but Poornima is second from the left and Wu is fourth from the left in the first row. Photo courtesy of Diane Wu.

Trace Genomics produces soil microbiome–testing kits that use genomics and machine learning. The company provides analysis of the data collected by its kits for growers, its target customers. The company's founders, Poornima Parameswaran and Diane Wu, are among a new generation of women developing solutions to some of the biggest challenges in agtech, historically a field dominated by men. When Wu and Parameswaran founded Trace Genomics in 2015, they barely understood the importance of soil health to farmers' businesses, but in the years since have become experts in the field.

As of 2020 Trace Genomics had some thirty-five full-time staff, including experts in biology, agronomy, data science, and software engineering. The company is still headed by Poornima Parameswaran and her business partner and cofounder, Diane Wu, along with company CEO Dan Vradenburg, former president of Wilbur-Ellis's agribusiness division. Outside its offices in the San Francisco Bay Area and Ames, Iowa, Trace Genomics maintains a sales presence at the Western Growers Center for Innovation and Technology in Salinas. So far, it has raised $22 million in total, more than the typical amount of capital raised by agtech start-ups.

Because of a widespread overall deterioration of soil health that has occurred on a global scale, farmers across the country—both conventional and organic—are accelerating efforts to achieve quality soil. Many are wrestling with the reality that quality soil has been depleted because of pesticides and traditional ways of farming such as tilling. One-third of the world's arable land has been lost due to erosion or pollution over the past forty years—with potentially disastrous consequences as global demand for food soars. Moreover, disease in soil can lead to crop failures and significantly impact the bottom line.[1]

As a result, growers are increasingly adopting techniques such as "cover cropping" and incorporating plants and livestock onto the farm to achieve better soil. Cover cropping is the technique of growing a mix of plants to cover the soil; this boosts soil quality and also manages pests, weeds, and diseases.

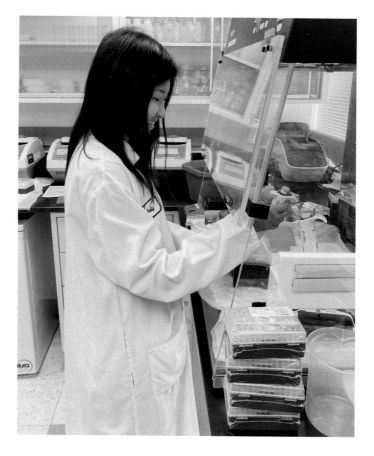

Wu working in Andrew Fire's laboratory at Stanford University. Photo courtesy of Diane Wu.

Trace Genomics creates soil-testing kits and uses DNA sequencing and machine learning to measure the health of soil, detect agricultural diseases, and help farmers maximize their yields. Machine learning uses algorithms to parse data, learn from it, and make determinations without human intervention.

With Trace Genomics' products and services, growers can compare soil health across multiple farms, manage disease risk by quantifying pathogen levels, and compare fertilizer performance and soil treatments.

Parameswaran and Wu don't come from agriculture backgrounds, and neither has a family history in farming. Wu is the daughter of a mechanical engineer and a chemist. Parameswaran's father was a petrochemical engineer and her mother a homemaker.

Wu majored in computer science at Simon Fraser University in Vancouver, British Columbia, and Parameswaran came to the US from Bahrain (where her father was stationed for work) and

Poornima and her parents and brother Prabhu at her graduation from Stanford University. Photo courtesy of Poornima Parameswaran.

earned a molecular biology degree at the University of Texas in Austin. Driven by a love for research and a desire to use her science skills in public health, Parameswaran enrolled in the PhD program in Stanford University's Department of Microbiology and Immunology in 2004.

"I was moving toward a career in the academy and going up the tenure track," Parameswaran recalls, "but once I got into Stanford I was bitten by the bug of innovation and really making an impact on society."

The women met at Andrew Fire's lab at Stanford University. Fire is a 2006 Nobel Laureate in physiology and medicine and served as both Wu's and Parameswaran's dissertation advisor. The two women discovered they shared a passion for using genomics and big data to tackle the roots of disease and applying their research to the real world. After earning her PhD in genetics and specializing in computational biology at age twenty-five, Wu worked for Palantir Technologies as part of a data science team dedicated to solving problems in areas such as fraud and crime.

Interested in starting her own company, Parameswaran attended the summer entrepreneurship institute at Stanford's business school. After graduating from Stanford in 2010, she went on to do postdoctorate work at the University of California, Berkeley, and while there worked with Nicaragua's health ministry.

The two young scientists started work together to develop a product to detect diseases in soil. They regularly drove down to the Salinas Valley, California, where they began connecting with farmers.

In 2015 they decided to focus solely on their fledgling start-up, which they named Trace Genomics. The company saw early success by winning spots in highly competitive accelerators—including THRIVE, an agtech

Poornima as a young child with her parents in Chennai, India. Photo courtesy of Poornima Parameswaran.

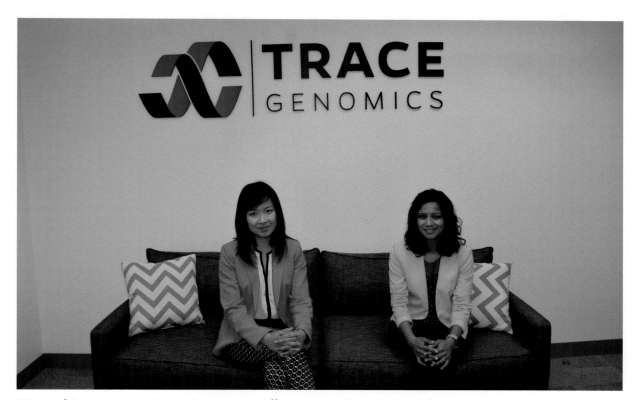

Wu and Poornima at Trace Genomics' offices in Burlingame, California. Photo courtesy of Benji Hsu.

Photo courtesy of Poornima Parameswaran.

accelerator based in Silicon Valley. At the end of the program Trace had beat nine other start-ups in THRIVE and was recognized as the most promising agtech start-up.

When it launched to market in 2016, the company started by offering a disease diagnostic tool for lettuce and berries. Since then, it has expanded its diagnostic coverage to over two dozen crops and now offers a comprehensive soil biology and chemistry diagnostic for soil.

These recent milestones in both raising capital and building a team are signs that Trace is on the upward trajectory of success. That said, since the company launched, the agtech space has also expanded with more players and investors. What is certain is that Wu and Parameswaran have blazed a trail as pioneers in a flourishing yet fast-growing sector and proved that age, race, and gender should not be barriers to success.

Soil health will continue to be a top priority for both conventional and organic growers. The industry's increased adoption of practices such as sequestrating carbon, incorporating production with livestock, encouraging the growth of native meadows, examining the combination of ecology and agriculture, reducing tilling, and experimenting with innovative growing methods such as cover cropping, rolling, and crimping are all signs that Trace Genomics is part of a greater solution to challenges posed by climate change and other factors.

Both women agree that keeping their corporate eye on their mission to solve problems for farmers is the key to success. For Wu, this means "harnessing technology to uncover the mysteries of the soil microbiome [a matrix that includes the bacteria, archaea, viruses, fungi, and protozoa that are found within soil] for the benefit of our global food system." And for her counterpart, Parameswaran, it means continuing to hone the "team we've built to tackle one of the biggest challenges facing our world today—the soil health of every arable acre." And for that there is a long way to go.

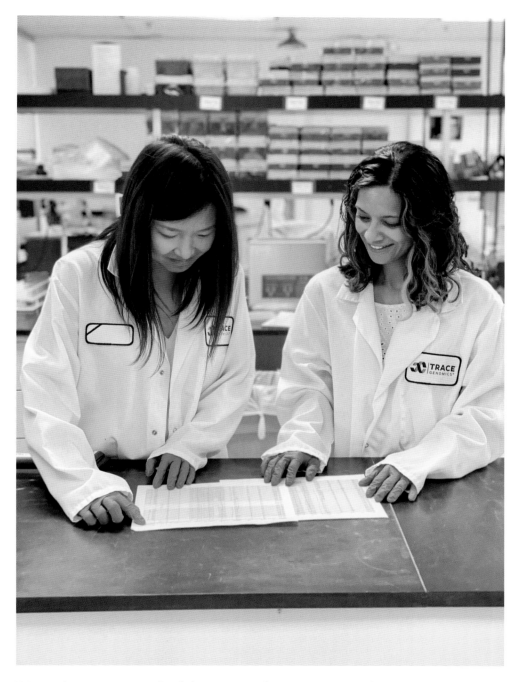

Wu and Poornima at the laboratory. Photo courtesy of Trace Genomics.

This profile is based on an article published in a series about female innovators in agtech in Techonomy *in January 2020, a digital publication that focuses on technology in society.*

Citations

Amy Wu, "Challenges and Opportunities: Diane Wu and Poornima Parameswaran," *Salinas Californian*, February 16, 2017, https://www.thecalifornian.com/story/news/2017/02/16/challenges-and-opportunities-diane-wu -and-poornima-parameswaran/97983584.

———, "These Female Entrepreneurs Grew Opportunity—in Soil," *Techonomy*, January 2020, https://techonomy.com/2020/01/these-female-entrepreneurs-grew-opportunity-in-soil.

Note

1. Grantham Centre, University of Sheffield, "Soil Loss: An Unfolding Global Disaster," December 2, 2015, http://grantham.sheffield.ac.uk/soil-loss-an-unfolding-global-disaster.

JESSICA GONZALEZ

Happy Organics, Merced, California

*A Focus on Bringing Fresh Local Produce
to Underserved Communities*

Jessica Gonzalez. Photo courtesy of Dexter Farm.

"Handle a book as a bee does a flower, extract its sweetness but do not damage it."

—John Muir

Happy Organics is a small batch beekeeper that provides raw honey and wellness products from some 34 hives. The company, launched in 2018 by Jessica Gonzalez, makes and sells its own products, including cannabidiol (CBD), lip and muscle balm, and CBD-infused honey. The company is based in the middle of the San Joaquin Valley, on Gonzalez's family farm, not far from downtown Merced. Happy Organics' mission centers on community, conservation, and wellness, as well as creating sustainable and delicious products.

Happy Organics is Gonzalez's latest venture in agtech. Previously, she was cofounder and chief technology officer of HeavyConnect, a Salinas–based agtech startup. In 2017 she left HeavyConnect and returned to her hometown of Merced after her mother died from illness at age sixty-nine.

Entrepreneurship appears to run in her blood. Gonzalez is the daughter of Mexican immigrants and the youngest of nine children, including three sisters and

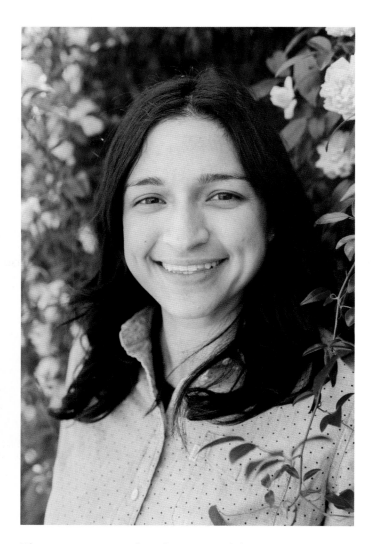

Photo courtesy of Melanie Smelcher.

five brothers. Her parents, Salvador and Angela Gonzalez, came to the US from Metrocan, Mexico, seeking a better life. The couple worked as farmworkers in the Salinas Valley and took on odd jobs

Gonzalez and her sister Rocio in the hoop house at the family farm in Merced. Photo courtesy of Amy Wu.

to make ends meet. In Salinas, Salvador started by selling fruit on street corners and expanded into wholesale. When Jessica was four, Salvador took a gamble and moved the family to Dinuba, outside Fresno, to expand his own company.

As a teenager, Gonzalez was a top student, gravitating to science and math. She was a valedictorian at Golden Valley High School and received a full scholarship to attend Mills College in Oakland. But during college she struggled with what she wanted to do with her life, especially after deciding the premed route wasn't for her.

"I really had no direction. I was just trying to figure out what should I be doing and what would be best for me," she says. An older sister, Sonia, encouraged her to look at computer science. "She said it's really marketable."

This turned out to be true. Even before graduating from Mills, Gonzalez received job offers from numerous Bay Area companies. In the end she accepted an offer from global technology company ThoughtWorks as a quality assurance analyst.

Discovering Agtech

It was ThoughtWorks that led Gonzalez to agtech and specifically to HeavyConnect. Gonzalez met Patrick Zelaya, a former employee at John Deere who had launched HeavyConnect, after she signed up for the mentorship program at ThoughtWorks. The mentoring program offered Gonzalez a list of university programs to select from, including CSin3 (a program that allows students to attain a degree in computer science in three years from Hartnell College and Cal State Monterey Bay).

Through the mentoring program Gonzalez met Rivka Garcia, then a student in the CSin3 program and an intern at HeavyConnect. In mentoring Garcia, Gonzalez became connected with Zelaya, who at the time was a one-man band with several interns. He was also looking to expand his company.

The decision to join HeavyConnect full-time was easy, Gonzalez says. She was enticed by the opportunity to experience working in a fledgling sector and

Gonzalez cleans old wax off wooden frames to reuse in the bee boxes. Photo courtesy of Melanie Smelcher.

to have a chance to develop a company from the ground up. She also liked the energy and buzz inside the Western Growers Center for Innovation Technology (WGCIT), where HeavyConnect was based at the time.

"They [WGCIT] brought a lot of different companies all together in one space, and I saw the different technologies being built. I was really excited about the possibilities," she recalls.

When Zelaya offered Gonzalez the position as cofounder and chief technology officer, she immediately said yes. The position gave her the chance to innovate and "have more control over what I am working on, and being at a higher level of deciding what happens with a company," she recalls.

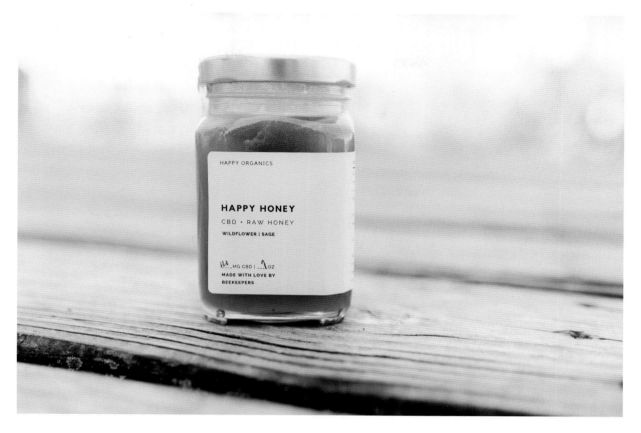

Happy Organics' flagship product is raw honey blended with CBD. Photo courtesy of Melanie Smelcher.

Once she joined HeavyConnect full-time in 2015, she also became aware of the realities of being a decision maker in a developing sector that was mostly male dominated.

"Initially, I was intimidated [when] going to the conference rooms where I am the only woman, but I think everyone is really receptive to what we are doing, so it's like they don't really see that difference," she says. "Every time we go into a conference room, they listen, they say we love this product, we need it."

Silver Linings

In 2017 Gonzalez left the startup and Salinas and returned to the family farm in Merced.

"I left after my mom passed away because I felt like I needed to find my purpose. I went back to the family business to see what I could help with and maybe help inspire a new idea," Gonzalez says.

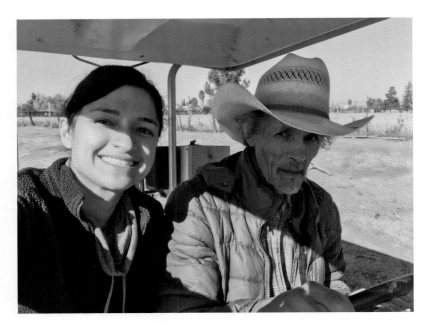

Gonzalez and her late father, Salvador, at the family farm. Photo courtesy of Jessica Gonzalez.

Later that year she was hit with more bad news. Soon after her mother passed away, her father was diagnosed with terminal cancer. She and her siblings helped their father with treatment, along with running the family farm. The genesis of Happy Organics started when Salvador found taking CBD helped with pain, but the oil in its raw version, which has a higher concentration of the plant material, was bitter to the taste.

"I thought mixing the two [CBD and honey] would kind of mask the cannabis taste and smell and I found people who had never tried cannabis were very interested in the honey," she says. She also made other products, such as CBD-infused oil tincture, muscle balm, and massage oil for her father. She found that the mixture helped her

Gonzalez, her siblings, and their spouses and children at their parents' anniversary celebrated at a rented ranch venue in Merced. Photo courtesy of Jessica Gonzalez.

32

Gonzalez at the family farm in Merced. Photo courtesy of Amy Wu.

father immensely with his pain as he went through medical treatments. His positive responses inspired her to take these products to the market.

Entrepreneurial in nature, Gonzalez began thinking of ways she could bring her passion for farming and technology together. At the family farm she started beekeeping, producing raw honey, experimenting with different recipes different recipes that blended CBD with honey, and making CBD-infused tinctures with the goal of taking the products to market.

At times, though, she still missed working in the agtech sector. The decision to leave HeavyConnect was difficult since she loved her job, but it was a necessary one. Her parents had taught the Gonzalez children that family comes first. When her father became sick, she and her siblings immediately regrouped and worked together to run the farm.

Turn the clock forward: in October 2018 Salvador Gonzalez succumbed to cancer. For Gonzalez, her personal losses have brought a silver lining: "I felt like I needed to find something I was really passion[ate] about. After a parent passes, you have these life-changing moments. I thought: life is short and I need to find something to do."

That's when she officially founded Happy Organics with the goal of taking the products to markets across the United States. The company is based out of the ten-acre family farm, which houses the beehives and a number of flower and fruit gardens where the bees collect nectar and pollen.

Happy Organics continues to gain momentum with customers as Gonzalez's story has been reported by the media. She is committed to making the honey-based products "as long as it keeps

Gonzalez brushing a bee swarm into a new box. Photo courtesy of Melanie Smelcher.

helping other people." Sales are inching upward with roughly 250 jars a month, mostly via online and at farmers markets.

In 2019 she launched a farm stand for a season that is now open every Sunday. She sells fresh seasonal produce, including leafy greens, zucchini, and tomatoes, to the community. "There's a lot of farmland on this side of town, but it's not easy to get access to what is being grown," Gonzalez says.

As of April 2020 Gonzalez was also working with Communities United for Restorative Youth Justice (CURYJ), a nonprofit based in Oakland, to build out its community garden. She plans to help CURYJ set up beehives and install a smart garden, which will help it automate and keep track of metrics for the garden.

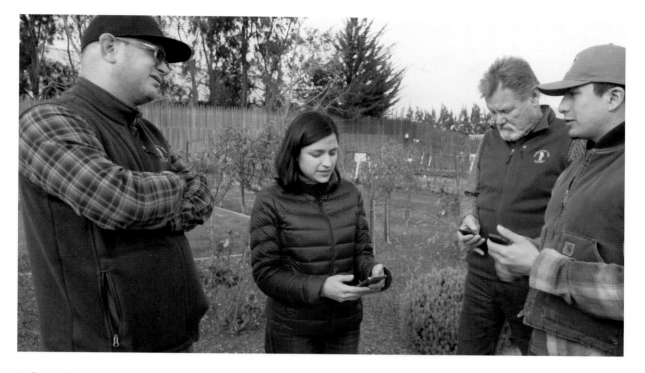

When she worked at HeavyConnect, Gonzalez demonstrated technology to growers. Photo courtesy of *From Farms to Incubators.*

"My parents always taught us to be respectful of nature and to care for the things around us. Conservation is our biggest priority and the quote reminds me of this," Gonzalez says, speaking of the John Muir quotation found at the beginning of the chapter.

Merced is home, but she also admits to having the itch to return to agtech, as a platform to develop innovation and find ways to help the community.

"I think agriculture is more meaningful for me—it's getting healthy food to lower-income areas and it's something I am passionate about," she says.

Citations

Jovana Lara and Jessica Dominguez, "Young Beekeeper Creates CBD-infused 'Happy Honey' Products for Sick Father," *ABC News*, May 21, 2019, https://abc7.com/mexican-american-beekeeper-merced-happy-organics/5309345.

Amy Wu, "Challenges and Opportunities: Jessica Gonzalez and Rivka Garcia," *Salinas Californian*, February 16, 2018, www.thecalifornian.com/story/news/2017/02/16/challenges-and-opportunities-jessica-gonzalez-and-rivka-garcia/97983794.

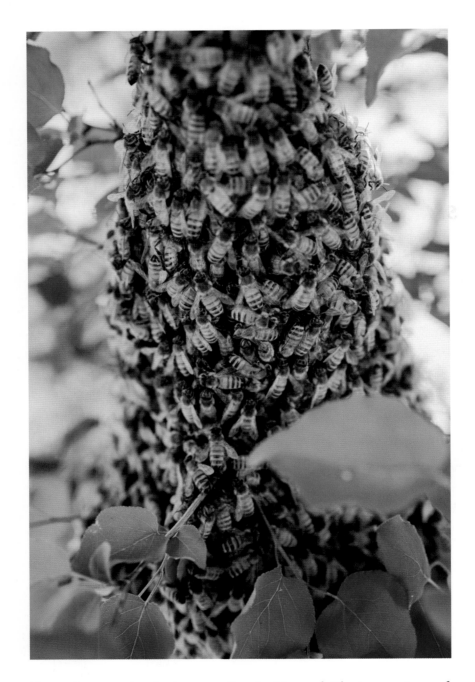

Honeybees at Jessica's operation in Merced. Photo courtesy of Melanie Smelcher.

MIKU JHA

AgShift, Santa Clara, California

This Entrepreneur Is Bringing Innovation to Food Inspection

Miku Jha. Photo courtesy of Miku Jha.

"If we can really understand the problem, the answer will come out of it, because the answer is not separate from the problem."

—Jiddu Krishnamurti

Late one autumn morning in 2018 in an office park in Santa Clara, California, Miku Jha met with her team of engineers. Jha's company, AgShift, is an artificial intelligence–enabled food-quality inspection platform that aims to offer better, faster, and objective quality assessments of produce and provide an alternative to traditional manual food inspection processes. Through a suite of technologies, including machines and mobile apps, AgShift has a mission to reducee food waste.

Jha and her team were refining the company's main product, known as the "Hydra," a machine roughly the size of a small refrigerator. The Hydra is a food-quality inspection platform that uses artificial intelligence (AI) algorithms and computer vision (a field of study that seeks to develop techniques to help computers understand the content of digital images such as photographs and videos). The platform inspects the quality of fruits and nuts and creates a faster, more accurate and efficient assessment system. Jha tells customers and potential investors that the technology takes the inspection process from an average of six to eight minutes to twenty seconds.

AgShift focuses on analyzing the quality of berries, nuts, and other high-value produce. Photo courtesy of Amy Wu.

Jha speaks at the Forbes AgTech Summit in Salinas. Photo courtesy of Miku Jha.

During the meeting, an engineer retrieved a tray from the machine and placed strawberries on it. Instantly, a high-definition monitor showed an image of the strawberries, which were then scanned for their quality.

AgShift's Hydra machine uses artificial intelligence to automate the quality inspection process for commodities such as strawberries and cashews. It collects data on every facet of the crop, from the color to contours, size, and ripeness, and uses neural networks (networks of artificial neurons that use algorithms to solve problems) to do the grading. The package also comes with a download-able mobile app that allows companies to easily capture and track inspection results.

Hydra has been actively tested at Driscoll's, one of the largest berry companies in the world, and at Olam International, a Singapore-based food company. To finance its development, AgShift has so far raised $6.5 million from two venture capital funds.

Farming Roots

Jha's journey into agtech certainly wasn't planned. She grew up on a farm in northern India, part of the fourth generation of a family that grows mangos. In 1996 she was one of the first female computer science majors at the University of Mumbai, eventually immigrating to the US to pursue a career in technology. She launched AgShift in 2016, with the goal of using technology to innovate safety processes in the food supply chain. The global food safety–testing market's value is projected to grow from $16.3 billion in 2019 to $23.4 billion by 2024,[1] driven in part by a growing middle class in the developing world. Jha's LinkedIn profile reads: "Tackling food waste head-on."

"Before AgShift, I was pretty blue," Jha says, referring jokingly to the time she spent before AgShift at IBM. At Big Blue, the nickname for IBM, while building various mobile-first products for the internet of things (a matrix of physical devices on closed internet connections that communicate with each other), she became curious about how sensors could be useful in agriculture.

Although she didn't have any experience in farming, Jha already came with the experience of launching companies. She founded and ran three start-ups, including Worklight, a technology firm that manages mobile apps across different platforms. IBM acquired Worklight in 2012, and Jha stayed on to build and shape IBM's mobile-first portfolio of products. She also pursued an MBA at Cornell University in New York. In 2014, she relocated from New York City to Silicon Valley to work on technologies to help small- to medium-size farmers. She brought along her husband and young daughter.

Once in California, she drove many thousands of miles to meet with different growers to "get a sense of how farming communities and [the] ecosystem were similar or different compared to

Jha speaks at the *Economist*'s Canada Summit in 2017. Photo courtesy of *The Economist*.

the small-farmholder community I grew up with, back in India," Jha says. The core question that drove AgShift's 2016 launch was, Can you bring technology and apply it to food challenges? She recalls herself asking, "Could vision technologies used for self-driving cars be extended to a machine designed for self-grading strawberries?"

Since entering the agtech space Jha has achieved several key milestones. In 2017 AgShift was chosen to join the THRIVE Accelerator. She has served as a panelist or speaker at high-profile conferences, including the *Economist's* Canada Summit and the Forbes AgTech Summit in Salinas, California. In March 2019 AgShift was the THRIVE-X Startup Challenge Winner, which came with a $100,000 prize.

Raising capital for AgShift has been more challenging than for her previous ventures. Some of that could be attributable to the fact that agtech is still a nascent sector and to the concern among investors that farming remains heavily dependent on weather patterns.

Nevertheless, Jha is confident AI is the answer to some of agriculture's biggest problems—including its severe and growing labor shortage in the United States. Jha believes technology and AI can eliminate the errors and subjectivity (ranging from perception of texture to color to size) that are inevitable factors of human inspection. "What you're trying to solve is very subjective, in the first place," Jha says. "If you say, 'Give me a count of all of these blueberries,' that's objective. That is very straightforward. But how do you measure AI success when tackling these [subjective] kinds

The Hydra is the innovation that provides the autonomous food inspection system. Photo courtesy of AgShift.

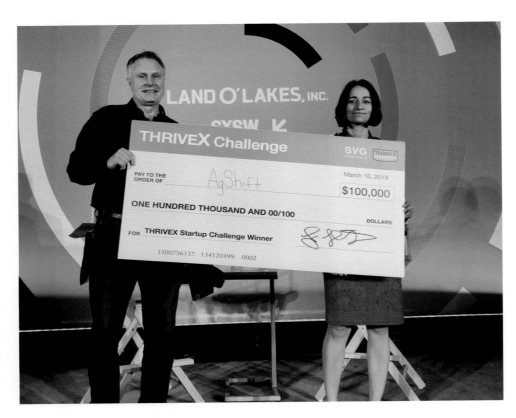

Jha accepts the award as the THRIVE-X Startup Challenge Winner in 2019. Photo courtesy of SVG Ventures.

of problems? Machines and automation replace the subjective with the objective, because software only has one eye."

One of Jha's main goals is to commercialize AgShift's core product, the Hydra, and expand its customer base beyond Driscoll's and Olam International. To better position the company to handle its growing customer base, Jha signed a deal to partner with a Pennsylvania robotics company that will produce some of the company's machines. AgShift's focus will remain on "high margin, high volume, high commodities," and there are plans to expand into produce like pistachios and a variety of berries, almonds, and avocados. "We will start exploring what sort of commodities match AI well, and invest research and development into new commodities," she says.

Weathering a Storm

In January 2020 AgShift moved to an office in San Jose near the San Jose International Airport. As of April 2020 AgShift had six full-time staff, in addition to founder and CEO Jah.

The team was on track to make the jump from the testing phase of the Hydra product to commercializing the product. The new Hydra was already being manufactured in Pennsylvania, and the company was ready to deploy it to customers, including Driscoll's and Wish Farms in Florida.

Jha with the Hydra in 2018. Photo courtesy of Amy Wu.

Then in April everything changed with the COVID-19 crisis. There was clearly a need for automation for food quality, considering that the volume of inspections hadn't changed, but the pandemic set the start-up back when it came to commercialization.

"Organizations are waking up—they need industrial grade automation. . . . The realization has been expedited because of the circumstances. There is a need to do better quality assessment. It's almost becoming an essential task," Jha says. That said, the pandemic hit just before commercialization and impacted the company's customers.

"If we were one season ahead, then organizations would significantly benefit. This quarter we were close to converting it [the Hydra] into deployment," Jha says. "We just got stuck in that one last mile." AgShift had planned another round of fundraising before launching the technology commercially in early spring 2020, but it had to put those plans on hold due to COVID-19 as potential customers found their businesses impacted both financially and operationally.

On the bright side, the pandemic has also shown that automation may not be an alternative but a necessity to food and farming in the near future. The upshot, Jha says, is that "for us, now it's much clearer that we are absolutely on the right track to what we are offering the food supply chain, because you can't depend on manual processes. The goal is how quickly can we support many more commodities for any scenarios in the future."

Jha speaks at the *Economist*'s Canada Summit in 2017. Photo courtesy of *The Economist*.

Jha believes innovation in technology doesn't eliminate the need for manual workforce but rather augments the existing skills and knowledge. Businesses are embracing the idea of AI across several components of their quality processes, and AI is a natural progression. Agricultural products are, by nature, dynamic. "We see differences within each crop season," says Jha, "and having a technology solution that can learn and adapt to these dynamic characteristics in real time is very appealing. Automation also helps to maximize employee resources. Taking a skilled but repetitive task out of an inspector's hands allows business to free up and cultivate that employee's talents to maximize her workplace environment. We do that by allowing her to manage a QC process not be the QC process."

Jha says she's committed to the start-up and weathering the storm. AgShift will continue to focus its technological road map on evaluating and inspecting fresh produce, with a particular eye on almonds and strawberries. The company will use this time to look for ways to improve its current product too and to eventually make the technology available to small producers.

Despite the setback caused by the pandemic, Jha remains focused on the goal of reducing food waste. There won't be any other start-ups on the horizon until AgStart is off and running and that goal is achieved, she says matter-of-factly.

Citations

Dick Anderson, "Technology Meets the Eye," Cornell University, September 13, 2019, https://business.cornell.edu/hub/2019/09/13/technology-meets-the-eye.

Pragati Varma, "Remove Biases from Food Quality Inspections with AI," Dell Technologies, May 22, 2019, https://delltechnologies.com/en-us/perspectives/ai-helps-standardize-food-inspection-to-reduce-food-wastage.

Amy Wu, "Challenges and Opportunities: Diane Wu and Poornima Parameswaran," *Salinas Californian*, February 16, 2017, https://www.thecalifornian.com/story/news/2017/02/16/challenges-and-opportunities-diane-wu-and-poornima-parameswaran/97983584.

———, "This Entrepreneur Is Bringing Innovation to Food Inspection," *Techonomy*, November 5, 2019, https://techonomy.com/2019/11/this-entrepreneur-is-bringing-innovation-to-food-inspection/?section=video.

Note

1. www.bccresearch.com/market-research/food-and-beverage/food-safety-testing-technologies-markets-report.html

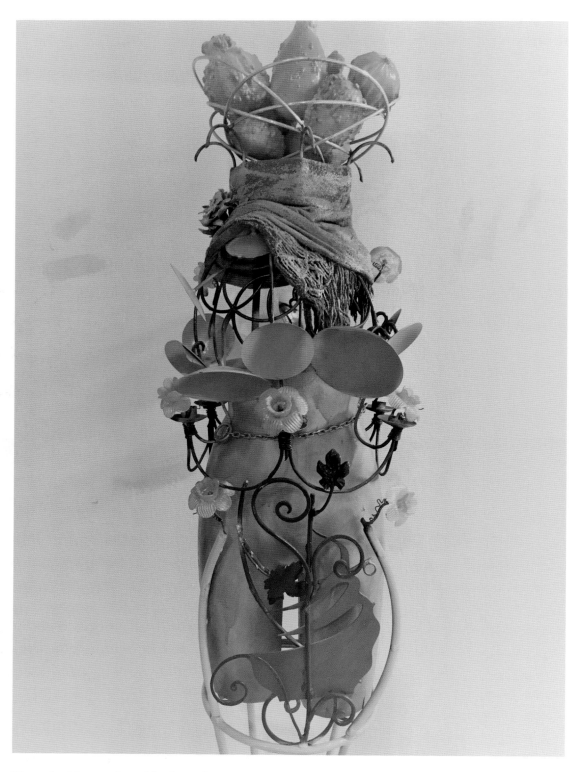

Thuy-Le Vuong (profiled on the next page) created a sculpture inspired by the theme of celebrating women in agtech.

THUY-LE VUONG

The Redmelon Company, Davis, California

A One-Woman Start-up Extracts Nutrient-High Oil from a Relatively Unknown Fruit

Thuy-Le Vuong with a redmelon at her sister's home in Southern California. Photo courtesy of Dan Chamberlain.

"Let food be thy medicine and medicine be thy food."

—Hippocrates

A matrix of emerald-colored vines cascade down the front of the ranch-style home in Elk Grove, a suburb of Sacramento. The leaves are palm sized and heart shaped. The woman who stands under the matrix is hopeful the vines will transform into a harvest of the fruit she calls redmelon.

Meet Thuy-Le Vuong (also known as "Le"), a scientist and researcher with a passion for connecting food and healing. Over the past two decades that passion has been focused on redmelon, her trademarked name for the fruit and also the name of her company. Called gac in Vuong's native Vietnam, the tropical fruit has oils that are rich in nutrition-packed carotenoids. Coral hued, the size of a bocce ball, and with thick, gnarly skin, the fruit is primarily found and grown in Southeast Asia, where it thrives in the warmer climate.

In 2016 Vuong officially launched The Redmelon Company with the motto "Nature at Its Best." For Vuong, the journey from the genesis of an idea to the launch of her company has been two decades and counting. Vuong developed an extraction method that retains the nutrients from the redmelon fruit while avoiding the use of organic solvents. The process involves a cold press, similar to an olive oil press, used to extract the solvents. Her overarching goal is to increase the fruit's accessibility for underdeveloped countries where the solvent isn't readily available. Vuong also sources and grows the fruit and commercializes the oil by creating products such as lip balm and nutritional supplements.

Photo courtesy of Dan Chamberlain.

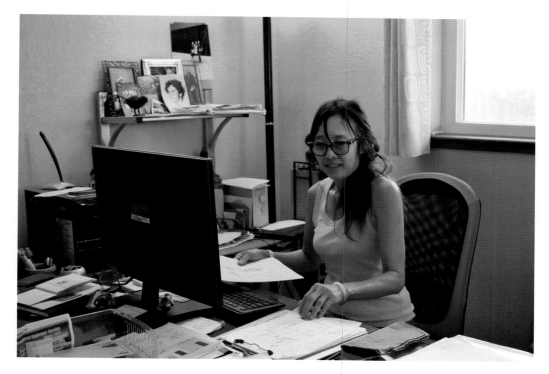

Vuong busy in her home office in 2018. Photo courtesy of Amy Wu.

Food and Culture

Vuong's fascination with food started as a girl growing up in Vietnam, much of it driven by "witnessing firsthand the cost of malnutrition." Vitamin A deficiency, which leads to immune deficiency and also night blindness, continues to be a significant problem in rural parts of Vietnam. A World Bank report found that one in three children from ethnic minorities in Vietnam suffer from stunting, while one in five are underweight.[1]

Vuong considers herself one of the lucky ones. Having grown up in an upper-middle-class family, access to food was never a concern, and she was surrounded by a wide array of flora, fauna, and animals.

"I always liked plants and flowers and enjoy experimenting with foods. Growing up in Vietnam, surrounded by so many different types of fruits, vegetables, plants, and animals, I was fascinated with how life can be maintained properly just by knowing and having the right foods. I strongly believe in the idea of let food be your medicine," she said. She further explained how the mission behind her passion comes from her parents.

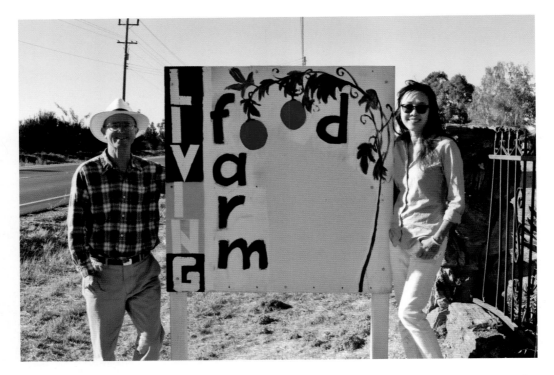

David Bersohn and Le at their former house in Elk Grove, which they named "Living Food Farm." Photo courtesy of Amy Wu.

Born and raised in Saigon, her father, Duc Vuong, was a teacher and later worked for the US Information Service during the Vietnam War. He loves gardening. At ninety he continues to "love planting and is a bonsai enthusiast."

Vuong's mother, now eighty-seven, was a business owner who launched a micro-credit group to help other women to obtain capital for their businesses. She also owned a bookstore. The family came to the US in 1975 at the end of the Vietnam War. Hungry to be a part of a Vietnamese community, they settled in Southern California.

As with many immigrant families, Vuong's parents were hoping their children would choose careers in medicine and engineering for stability. Her father tried to talk her into attending medical school, but with her love for data she instead gravitated toward computer science. In 1979 she earned her bachelor's degree in computer science from California Polytechnic State University and later an MBA from California Lutheran University.

Vuong worked for fifteen years as a computer scientist, including for Rockwell International and a variety of start-ups, before pursuing her PhD.

Founding Redmelon

Vuong says she was familiar with gac from her childhood in Vietnam. She recalls it was often eaten during the holidays, its juices and oils—the color of blazing fire—used to dye rice in festive occasions.

She didn't start researching the fruit until she pursued her PhD in nutrition at University of California, Davis in the early '90s. Her love for researching the exotic fruit was driven by a fascination with food as a means of healing. Her focus on gac's health benefits was also inspired in part by her daughter's medical health problems. Vuong's daughter was born prematurely and had substantial metabolic issues that made digesting milk challenging. As a young mother, Vuong sought alternative foods with nutritional value for her daughter. The passion to focus on foods that boost health and wellness remained with her when she joined Davis.

At Davis she took a deeper dive into gac and discovered the fruit contained oil with substantial nutritional value. In her research and reading she confirmed that redmelon—specifically the fruit's seed membrane—contains oil, making the lipophylic nutrients extracted from the oil more stable and bioavailable.[2]

In 1997, when she was at UC Davis, she conducted the "Xoi Gac Supplementation

The inside of a redmelon. Photo courtesy of Thuy-Le Vuong.

Trial" as part of her PhD dissertation. The study was carried out in seven villages in northern Vietnam where some three hundred pre-school-aged children participated. Over a thirty-day period the children were fed one of the following meals: rice colored by redmelon (which contains a high concentration of beta-carotene), rice colored by red food colorant, or rice colored by synthetic beta-carotene. The objective was to show that redmelon can improve the nutritional status of vitamin A–deficient children. The results showed that among the children in the redmelon group, vitamin A status improved, as did hemoglobin concentrations, the lack of which is the underlying problem. The improvement was significantly higher for this group than for the group receiving. synthetic beta-carotene, Vuong says.

She has been trying to increase the raw material needed to make the products (notably edible oil) by growing her own redmelons. Although the US Department of Agriculture (USDA) has banned the import of the fresh fruit into the US due to concerns over exotic pests, while at Davis she received a USDA permit to import a whole fruit

Vuong uses oil extracted from red-melons to produce Rosolive. Photo courtesy of Thuy-Le Vuong.

for research, and started growing redmelon in the university greenhouse to create a supply. She has since collected many seeds and plants from that initial fruit.

Vuong invented a non-solvent method to extract the beta-carotene-rich oil from the fruit. The method retains the nutrients of the oil and avoids using organic solvents, which are not readily accessible in underdeveloped countries. The end product of the method was the oil.

After earning a PhD in nutrition and epidemiology at UC Davis, she spent three years working as a research nutritionist at the USDA Western Human Nutrition Research Center in Hawaii before making the leap into start-ups.

Redmelons. Photo courtesy of Thuy-Le Vuong.

Discovering Redmelon

In 2011 Vuong launched Fishrock Laboratories to commercialize products made from gac oil such as capsules, lip balm, and edible oil and to make it assessible to general consumers. She referred to the name of the fruit as "redmelon," after the color of the fruit itself, and the fruit's oil as "redmelon oil." Later she trademarked the name "redmelon." She saw incorporating the oil into everyday products as a way to get it to the general public.

The products were received positively by friends and mentors, including J. Bruce German, professor and chemist of food science and technology at UC Davis, and Suzanne Murphy, a former UC Davis professor and director of the Nutrition Department at the Cancer Research Center in Honolulu, who mentored Vuong on her PhD thesis.

Vuong also gained the attention of Pam Marrone, founder of Marrone Bio Innovations, a Davis-based company that specializes in producing biopesticides. Marrone was attracted to redmelon for its nutritional value and its potential as a "new superfood."

In 2016 Vuong changed the name of the company from Fishrock Laboratories to The Redmelon Company. That same year, she ramped up the commercial side of the business with large-scale production. She collaborated with a large olive oil–producing facility in Hayward, California, and they produced an inaugural product called Rosolive, made of olives and redmelon pressed together. The product's main distribution channel was through farmers markets as well as through word of mouth. A year later, her company officially launched the brand Redmelon, with the motto "Nature at Its Best."

Building a Team

Over the past few years Vuong's focus has been on fundraising and building a team. She has knocked on the doors of traditional venture capitalists and met with Fortune 500 food-based company executives who listened to her pitches with interest but ultimately decided not to move forward.

"In a decade on my own, I have great ideas, patents, and supportive friends, but business success has been elusive," she says.

"Vuong doesn't have a team," says Marrone. "I have introduced her to investors, she has a patent on oil extraction, and she has products that work really well such as the skin cream and supplements. She has a really good product, [but] she needs a team. There are investors who said she's not a CEO type, but she needs to find someone who fits her values, and that is a big challenge. She really needs a CEO or COO to take it to the next level. That's very important; otherwise it's just you and the technology."

The twists and turns and curveballs—and frankly the hardships—of being an entrepreneur somehow balanced out with her life partner, David Bersohn, whom she had met on a bus ride.

David, a soft-spoken man in his sixties who looks about a decade younger than his age, is a musician and an artist (specializing in metal sculpting) who also balances art with day trading. He has accompanied Vuong to various conferences where she has showcased her product and innovations, and represented the company at exhibitions for her at the conferences. He has also invested in the company.

In 2017 the couple bought a five-acre horse farm in Elk Grove. They hung a shingle, "Living Food Farm," alongside Grant Line Road with the goal of experimenting with growing redmelons and expanding the supply of oil. She started growing the fruit to see if it could be cultivated in Northern California

The seed of *Momordica cochinchinensis* (gac or redmelon) plant is hand engraved with the motif interpreting the story of Thuy-Le Vuong and The Redmelon Company. Photo courtesy of Sergey Jivetin.

and thus bring a much-needed increase to its supply. By midwinter, however, it became apparent the melons that she planted would not mature beyond the vines and flowers. There was no fruit. Vuong says the area "wasn't optimal" because of the cold winter, which had days when it had dropped below zero. At the end of 2018 there were more turning points: she and David married, they sold the farm and moved to a much smaller home in Davis—and took the redmelon vines with them.

As of early 2019 the Redmelon venture appeared to have taken yet another turning point. After two years of discussions with Kristy Levings of United Crop Research, a young company that focuses on commercializing exotic fruit, there was a possibility that United Crop would acquire Redmelon. Says Levings, "Redmelon has a proven product that can be scaled up."

For a long time Vuong was protective of her company and never considered selling it, but that has changed since she has accepted that capital is critical to her dream. "I think it's always my interest to work with natural ingredients and food products," she said. Vuong continues to hold true to her vision: "I really believe redmelon is the next super fruit."

But entrepreneurship has its rapid twists and turns. As of 2020 Redmelon's talks with the folks at United Crop Research ended with Vuong saying their vision of redmelon did not align with

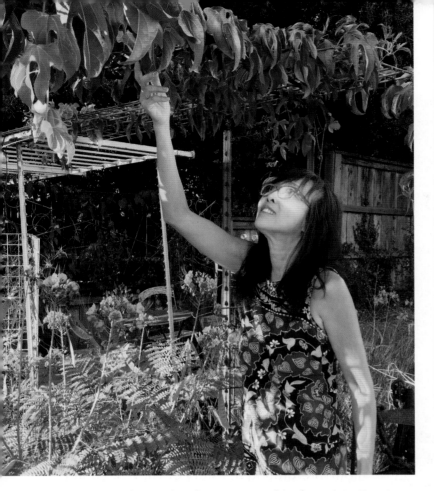

hers. Since then, Vuong has generated a small amount to produce more bottles of Rosolive with a plan to sell them at farmers markets; production was done at the Bozzano Olive Ranch mill in Stockton, California.

The fruits continued to grow at her sister's home in Southern California. She and David have settled into Davis, which they consider home. The house is framed with redmelon vines (as of spring 2020 they look promising, she says) and the inside shows an art studio with dozens of metal art pieces, one of which includes Vuong smiling and holding a redmelon. Here she and David hold out hope that the vines will bear fruit.

Vuong under redmelon vines at her home in Davis, California. Photo courtesy of Thuy-Le Vuong.

Citations

Amy Wu, "And Their Daughters after Them," *Salinas Californian*, February 17, 2017, https://www.thecalifornian.com/story/news/2017/02/17/and-their-daughters-after-them/97860768.

———, "Women in Agtech: From Vietnam to Founding Redmelon Company—Le Vuong,"*AgFunder News*, July 27, 2017, https://agfundernews.com/women-in-agtech-from-vietnam-to-founding-fishrock-labs-le-vuong.html.

Notes

1. World Bank, "Vietnam: Ethnic Minority Children Are Disproportionately Undernourished," press release, December 10, 2019, https://worldbank.org/en/news/press-release/2019/12/10/vietnam-ethnic-minority-children-are-disproportionately-undernourished.

2. Thuy-Le Vuong and J. C. King, "A Method of Preserving and Testing the Acceptability of Gac Fruit Oil, a Good Source of Beta-Carotene and Essential Food Acids," *Food and Nutrition Bulletin* 24, no. 2 (2003): 224–30, https://pubmed.ncbi.nlm.nih.gov/12891827/

"My early fieldwork of my research [on redmelons] was carried out in northern Vietnam. I was fascinated with the hard-working women and playful children in rural areas. Both paintings are recapturings of scenes in villages, mostly during the time the fruit is ripening. It is also the rice-harvesting time."
—Thuy-Le Vuong

Crimson Guild (above) depicts a scene of village women preparing redmelons for a traditional rice dish called *xoi gac*, or rice colored red by gac (redmelon) for New Year's celebration.

Fruit from Heaven (right) shows the courtyard of a village home, with a large vine of red-melons, a sack of rice, baskets of ripe fruits ready for the market, and children playing by the gate.

PAM MARRONE

Marrone Bio Innovations, Davis, California

A Pioneer in Agriculture and Technology and a Mentor for a New Generation

Pam Marrone at the research laboratory at Marrone Bio Innovations.
Photo courtesy Amy Wu.

"When you get into a tight place, and everything goes against you till it seems as if you couldn't hold on a minute longer, never give up then, for that's just the place and time that the tide'll turn."

—Harriet Beecher Stowe, abolitionist and writer

Trying to catch a moment with Pam Marrone is like trying to catch a rainbow. A fast talker with an infectious smile and a twinkle in her eyes, she is a successful entrepreneur and a self-made pioneer in the world of agtech. At sixty-three, Marrone has the energy of someone half her age, and her voice is clear and audible. It's hard to imagine that anyone would ever ask her to "Please speak louder."

Marrone Bio Innovations (MBI) was launched by Marrone in 2006 as a bio-based pest management company that produces herbicides, fungicides, and insecticides. The company's goal is to have its products—which are plant based or use naturally occurring microorganisms—replace traditional chemicals used in agriculture. In 2013 the Davis-based company went public on the NASDAQ with the MBII ticker symbol. It now has 130 employees, including many scientists and researchers. Marrone, in addition to her work in her own company, has long served as a mentor and advisor to women entrepreneurs in agtech and invested in a number of their companies.

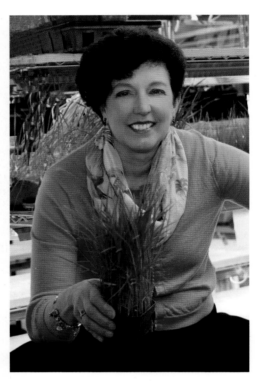

Photo courtesy of Pam Marrone.

After fourteen years of running Marrone Bio Innovations, she could retire. In fact, in 2019, when Marrone announced that she'd be stepping down as CEO, some newspapers in Yolo County, where Marrone Bio is headquartered, reported that she was retiring, which she now says isn't exactly accurate.

Marrone was quoted in the *Sacramento Business Journal* as saying: "This is the 30th year of me starting my first company in Davis, with no break in between. It would be nice to not be CEO of a publicly traded company, with the constant grind of quarterly earnings. . . . It's a complex job in a complex industry," she continued. "I am proud of where we got the company to."

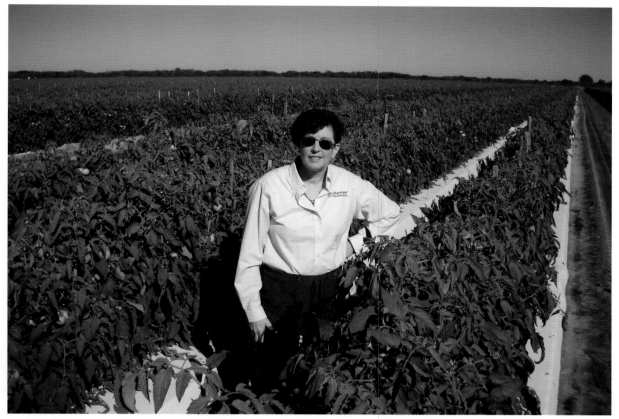

Marrone standing in a tomato field in Florida. Photo courtesy of Pam Marrone.

Yet by the start of the new year, at a time when she could clearly sit by a golden pond, Marrone wants it known she is not retiring. In fact, she had never planned on *retiring*. In January 2020 there she was at EcoFarm, one of the premier conferences for organic farming in the West Coast, held annually in California's Central Coast region. She was front and center at the tours and workshops, actively taking notes on her tablet and asking questions. Many of the farm's attendees visited were her customers, either already using her product or working with her on research trials. She studiously sent reports back to her company, and when she wasn't on the tours or workshops, she was found at Marrone Bio's exhibition booth chit chatting with attendees.

Marrone typically has a variety of projects in the works and the wheels are constantly turning. Another start-up is in the pipeline. "I am looking to start a new herbicide company and am looking to recruit younger women CEOs to run the company. I will be chairman," says Marrone. Although early in the works, she describes it is as a "grower-focused company [that] solves big problems. It focuses on what is the next thing that will be banned, and comes up with a solution sooner. It's going to be totally disruptive."

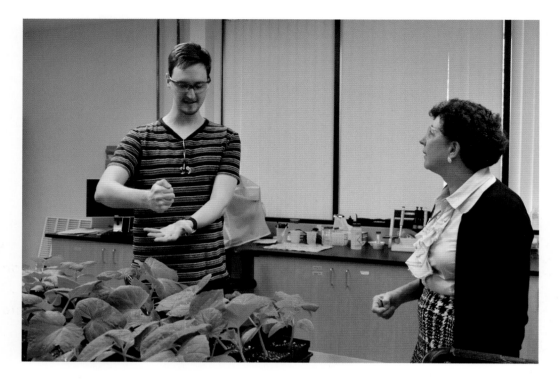

Marrone with a staff member at Marrone Bio Innovations. Photo courtesy of Amy Wu.

She is also consulting for Bob Woods, the chairman of the board of Marrone Bio. There is a twinkle in her eye and excitement in her voice when she talks about the company's products. Perhaps her enthusiasm goes back to her roots as a trained researcher and scientist in entomology, first as an undergraduate at Cornell University and then as a PhD at North Carolina State University.

MBI is in the process of commercializing a fungus that was found on the bark of cinnamon trees in a Honduran rain forest. MBI is growing the fungus on barley grain and disking (a soil preparation practice that typically follows plowing) it into the soil. "It puffs out gases before you plant the strawberries, and it fumigates the soil, but it doesn't harm any of the microbes," explains Marrone.

Javier Zamora, a Salinas Valley–based organic farmer who specializes in year-round production of strawberries, has been working with Marrone on a two-year trial of the fungus. The trial—which is funded by a grant that Marrone Bio earned through California's Healthy Soils Initiative—involves measuring the gas emissions after broccoli and romaine lettuce are grown.

Zamora says he is impressed with how responsive Marrone is. "That lady kicks ass. She's a hard worker and committed to doing beautiful things for the environment," he says. "She's very dynamic. I have a lot of respect for her. She's always working and she knows how busy I am, but she keeps chasing me. But she's not like a pain in the ass. She's very particular on what she wants. She does not take the easy way."

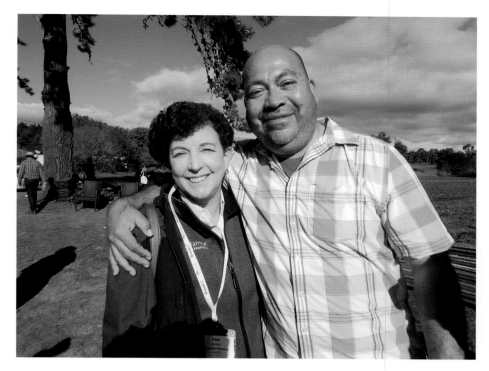

Marrone and Javier Zamora, founder and owner of JSM Organics. Photo courtesy of Amy Wu.

She's also in the early stages of setting up a nonprofit, a foundation that will fund products with naturally occurring microorganisms that tackle invasive species, such as mussels, pine bark beetles, and toxic algae. The nonprofit structure will allow for grants and also commercial success.

"So, as you can see, I have lots of irons in the fire," says Marrone.

From the Ground Up

The time to explore different initiatives and to some extent passions came with a price that Marrone has paid.

When I first met Marrone in 2017, she was in the throes of running Marrone Bio. A typical day started gangbusters at 6 a.m., when she'd get up to jog and walk her dogs. She is an avid runner and weightlifts during off days. Marrone equates exercise with a necessity and calls it "critical to mental sanity." This had been her routine for the past thirty-plus years.

Regularly, she'd eat breakfast at the company, often enjoying the product of one of her beloved hobbies—baking (on the day I went to visit at the end of 2018, it was banana bread). When not traveling (and she frequently did whether it was speaking at a conference or tending to her duties as board member at her Cornell University alma mater), she was a common sight at the office, walking through labs, greeting researchers, asking how certain projects were going, and refilling the

Marrone examining grape vines at a vineyard in Petaluma, California, in 2006. Photo courtesy of Pam Marrone.

chocolate bars in the staff kitchens and waiting areas. At company gatherings such as the employee soup fest she's known for making vichyssoise.

Marrone's high energy and outspokenness can bely the reality that although she is an entrepreneur, she is also very much a scientist. Plants and gardening were a part of her upbringing. She was born and raised on a forty-acre family farm in Connecticut. Her father had Italian roots and her mother, Polish roots. There were a total of five children.

Both parents were avid gardeners and grew most of the vegetables they consumed. Notably, Marrone's mother was adamant about growing food organically, and this had a lasting influence on Marrone.

Marrone attended Cornell University and majored in entomology with the goal of someday creating products that were grower friendly and green friendly. It was ultimately the jobs she held post-PhD that pushed her into starting her own company. She was dissuaded by the red tape at Monsanto, where she was employed from 1983 to 1990 as the head of the insect biology section. In 1990 she received an offer from Denmark-based Novo Nordisk to launch and lead Entotech Inc., which was eventually sold to Abbott Laboratories.

"I really liked the natural products and I missed it," says Marrone of her departure from Monsanto. Moreover, "I really wanted to have an impact in agriculture, and I was really repelled by the big-company political scene."

Marrone and her mother at the opening bell at the NASDAQ when Marrone Bio went public. Photo courtesy of Pam Marrone.

When it comes to entrepreneurship, the turning point came in 1995 when she launched her own start-up biologics company AgraQuest, a supplier of biological pest management. The venture was Marrone's first lesson in fundraising, in which she had little experience. This was in the pre-internet days when the tools boiled down to the National Venture Capital Association directory and cold calling, Marrone says.

She tackled every fundraising round of the company on her own, clicking off A, B, C, and D and onward. "I worked like hell," Marrone says. By 1995 she had raised $60 million for AgraQuest. After Marrone left, the company was sold to Bayer for $500 million.

In April 2006 she rented a lab and pulled together a team of founding members and invested money to establish Marrone Bio Innovations. In the seed round of funding she raised $230 million (the majority from venture capital), including $50 million of her own money as well as money from family and friends.

The company has seen its share of dark valleys, including being hit with a scandal. In 2013 chief operating officer Hector Absi was charged with securities fraud, changing the books for a performance bonus. MBI paid a $1.75 million settlement with the US Securities and Exchange Commission; lawsuits and legal settlements cost the company more. The company was forced to cut its workforce in half, from 161 in 2013 to 84 in November 2015 (it has since increased again to 130 employees).

Marrone connecting with former cecretary of agriculture Dan Glickman and Keith Pitts, Marrone Bio's chief sustainability officer and Senior vice president of government and regulatory affairs, at Foster Our Future 2020, the annual meeting of Foundation for Food and Agriculture Research Annual Meeting in 2020. Photo courtesy of Pam Marrone.

It was a difficult time for Marrone; working closely with an executive coach and the encouragement of her husband helped greatly. She had considered throwing in the towel (and has considered it many times), but each time something held her back.

That said, MBI products have been increasingly popular and have taken advantage of the overall growth of the organic agriculture market. The products are the output of a vast team of scientists who are charged with creating biofungicides that help growers with pest management. The company's first product was Serenade, which stopped mildew growth on vines. Four core products followed, including Regalia, a plant extract with fungicidal and plant health activity, and Zequanox, a microbial molluscicide for the control of invasive zebra and quagga mussels. Marrone herself also holds 250 patents.

Today there are 130 full-time staff, of which roughly 34% are in research and development.

Marrone and her mother Florence in Killingworth, Connecticut, at the Marrone pond. Photo courtesy of Pam Marrone.

Mentor and Advisor

Along the way Marrone hadn't expected to take on another role—that of serving as an unofficial mentor and advisor to women entrepreneurs in agtech. Over the years, many of these women have connected with her through conferences or events where she has spoken, and others have run into her because they share the same ag and tech circles in the Davis area. They will call her, asking for advice on everything from building a team to fundraising.

Marrone has taken on this role with enthusiasm and made it a point to mentor a new generation of women in agtech. "I open my Rolodex and introduce them [the women] to investors," Marrone said. "I tell them go to the Village Capital incubator. I'll read their business plan and help them. Sometimes I help them with a little money." In some cases, such as The Redmelon Company, Marrone has invested in the company (as of 2020 she was serving as its interim CEO of the company). In other cases, she has served as a board member. She helped entrepreneur Fatma Kaplan—the CEO and founder of Phernoym—connect with investors. She sits on the boards of four women-launched agtech companies: AgShift, Pheronym, Redmelon, and StemExpress.

"If I like the technology and the entrepreneur, I'll do it," she says matter-of-factly. In most cases she promotes their work, connecting the women with journalists and sharing their accolades through her social media network. Seeing the value of mentorship was part of her upbringing, as

Photo courtesy of Dexter Farm.

both her parents regularly volunteered, whether it was serving on the local school board or visiting the elderly.

Another element of Marrone's pay-it-forward playbook stems from the challenges she's faced as a female entrepreneur in the agbio and agtech spaces. Her PhD broke barriers: "They assumed I should be chief science officer," says Marrone. "I spent a lot of time breaking out of that."

Marrone continues: "Also, in business women have a much narrower range of acceptable behavior. . . . There's a lot of times when you'll say something and it won't be credible, but if guy says the same thing it will be credible. . . . Also there are powerful men who don't want to see me as equal."

At times the greatest bottlenecks came from her own board or her own staff. She half-jokingly says she was afflicted by "strong woman syndrome" and points out that raising her voice means getting typecast by other women as overbearing. "I find more women get offended if I tell them what to do or disagree," she explains.

Over the years she's amended her management style. "I pick my battles more carefully, and sometimes when I know someone is going down the wrong path, I will use other people to course-correct or call a meeting or get everyone to talk about it," she says.

On surviving and thriving as a female entrepreneur, Marrone observes: "You've got to be very resilient and you've got to have an armor so that rejection bounces off you. It's really a street fight when you're an entrepreneur. Not everyone has the personality to do it."

Finding Balance

Marrone considers herself lucky having found an equilibrium with her husband, Mick Rogers. They've been married for forty-two years and counting, having met each other at Cornell as undergraduates. He is a social worker with an MBA and recently earned a PhD in social work at Smith College. She regularly returns to the East Coast where she attends meetings as an alumni-elected trustee of Cornell University and visits her ninety-three-year-old mother. As founding chair of the

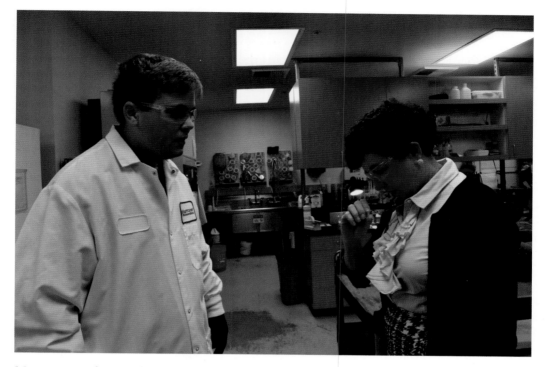

Marrone with a staff member at Marrone Bio Innovations. Photo courtesy of Amy Wu.

Biopesticide Products Industry Alliance (BPIA), a trade association of more than 150 biopesticide and related companies, she chaired meetings, led the group through strategy, and recruited staff from Marrone Bio Innovations to serve on the board and committees.

Reflecting on the past decades, Marrone thinks that things are changing". The agtech sector is gaining steam and funding as agriculture is forced to innovate to stay competitive. There's a new generation of women entrepreneurs moving into the agtech and agbio space. Often, they are younger—in their twenties and thirties—and they would rather start their own companies that work to solve problems in ag than work for a big conglomerate.

"People say I'd rather work on something I believe in, and there's a lot more infrastructure for entrepreneurs. They want to work in an area that would give them the satisfaction of making a difference," she observes.

Marrone's passion for mentoring the next generation continues. In February 2020 she could be found at UC Davis addressing a classroom of undergraduates about careers in agbiotech.

With cannabis and hemp fast emerging in the industry, Marrone sees even more opportunities to help growers. In early March 2020 Marrone was at the Biocontrols USA Conference in Portland, Oregon, to discuss the recent Environmental Protection Agency (EPA) approval for use of BMI's product Stargus on hemp.

When the COVID-19 crisis hit soon after, MBI received approval from the EPA to use its product Jet-Oxide (hydrogen peroxide plus peroxyacetic acid) to tackle coronaviruses to sanitize industrial food and agricultural hard surfaces. MBI owns the product as part of its acquisition of a small Florida-based company but did not expect that it would be used to fight the virus. "Serendipity. We didn't expect that at all," Marrone says.

MBI also launched Pacesetter, which increases corn, soybeans, and wheat yield. She asserts that problems such as labor shortages "have been exacerbated with COVID-19, so as a result I am seeing increased mechanization and farmers using agtech more and more precision tools, and things that will help them get a better return on investment." Marrone forecasts, "Overall, it looks like agriculture will weather this better than other industries because agriculture does have to produce food." The sky is the limit when it comes to both agriculture and innovation.

Citations

Mark Anderson, "Marrone Bio CEO Pam Marrone to Retire," *Sacramento Business Journal*, December 2, 2019, https://www.bizjournals.com/sacramento/news/2019/12/02/marrone-bio-ceo-pam-marrone-to-retire.html.

Dale Kassler and Denny Walsh, "Former Marrone Bio Executive Charged in Federal Fraud Case," *Sacramento Bee*, February 17, 2016, https://www.sacbee/new/business/article60842687.html.

Michael McGough, "Marrone Bio Innovations Founder Retiring as CEO, Davis-based Company," *Sacramento Bee*, December 2, 2019, https://www.sacbee.com/news/business/article237971314.html.

Amy Wu, "And Their Daughters after Them," *Salinas Californian*, February 17, 2017, https://www.thecalifornian.com/story/news/2017/02/17/and-their-daughters-after-them/97860768.

MARTHA MONTOYA

AgTools, Irvine, California

This Woman Entrepreneur Gives Farmers Information to Grow

Martha Montoya. Photo courtesy of Martha Montoya.

Montoya speaking about AgTools' technology during the AgTech X event sponsored by the Western Growers Center for Innovation and Technology. Photo courtesy of WGCIT.

"It is not the strongest or the most intelligent who will survive, but those who can best manage change."

–Charles Darwin

Martha Montoya never imagined that her journey from Bogotá, Colombia, to California would take her into a thriving agtech career. Montoya is founder and CEO of AgTools Inc., a food supply platform offering real-time news and information to farmers and agriculture buyers on over seventy-five different market variables, from weather to transportation to distribution to pricing for more than five hundred different commodities. The company's product aims to help growers, especially small farmers, manage market volatility, increase profitability, and reduce food waste in the supply chain.

Montoya compares AgTools' subscription-based SaaS (software as a service) platform—which can be accessed through mobile apps—to the Bloomberg terminal, a computer software system issued by Bloomberg LP that provides financial data. The information provided by the platform is curated by a team of AgTools' representatives and country managers, many of whom are scientists, researchers, and agronomists. Team members pull the data from a variety of sources, including the federal Departments of Agriculture, Transportation, and Defense. They pull similar data from analogous agencies and

Montoya demonstrates her technology. Photo courtesy of Martha Montoya.

departments in other countries. Montoya notes that all information supplied to AgTools is curated from official agencies. As of 2020 the company had fourteen employees throughout five offices in Colombia, Mexico and the United States.

Montoya is fast talking, fast moving, and dynamic and appears very quick to respond to rapid changes and challenges. She grew up in an upper-middle-class family in Colombia. Her father

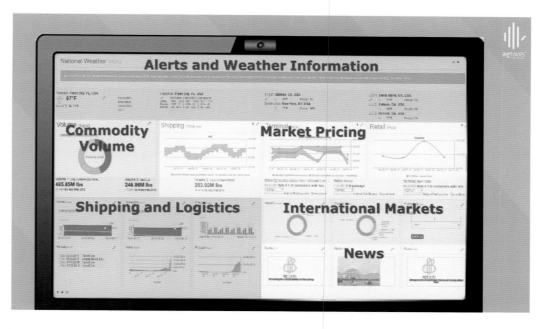

AgTools is a food supply SaaS (software as a service) platform that provides real-time intelligence to farmers and buyers with the goal of reducing food waste globally.

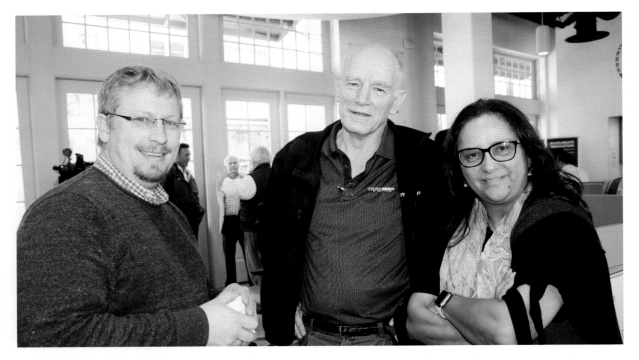

Montoya with fellow WGCIT start-ups leaders Nathan Dorn of Food Origins and Colin Brown of TracMap during WGCIT's three-year anniversary celebration. Photo courtesy of WGCIT.

founded the first night schools in Bogotá, offering continuing education for the capital city's working people. She recalls traveling easily between the city's relatively wealthy and impoverished neighborhoods, an experience that informs her current work. She had planned to follow her father's footsteps in education, and graduated with a degree in biology and chemistry from La Salle University in Colombia with plans to be a teacher. She was also a competitive tennis player during college and competed throughout South America in her youth.

With political strife and drug cartels wracking Colombia, Montoya immigrated to the US in 1989. At twenty-five and newly married, she initially passed through Los Angeles on her way elsewhere. Later, homesick for a community of Colombians, she returned to Los Angeles and settled in Orange County, California.

Montoya attributes her work ethic to her father, whose mantra was that no job assignment was out of the realm of possibility for her "as long as you work and they pay you." Even as she applied for jobs, she cleaned houses to make ends meet. Through classified ads she found a position at a school library.

In a bit of a twist, it was her passion for cartooning (at one point she wanted to be a full-time professional cartoonist) that led her to agriculture. She did find some success in cartooning and

In 2019 AgTools was a winner of the John G. Watson Quick Pitch Competition. Photo courtesy of UCI Beall Applied Innovation.

in 1995 licensed her cartoon characters in a comic strip named *Los Kitos*. At its peak the comic strip ran in over 200 newspapers internationally, including *La Opinión* in Southern California. But the gig was not enough for her bread and butter. She answered an advertisement from a trade company seeking a receptionist, figuring the job would afford her the time to pursue her cartooning outside of work.

In reality, the job extended far beyond answering phones and included buying supplies globally. The position utilized her knowledge of chemistry and language and her love for travel and working with folks from all over the world. Montoya built a vast network in the agricultural sector, especially in Latin America, where her ability to speak both English and Spanish was a great asset.

In 2010 she founded Los Kitos Produce and Farms (named after her cartoon strip) in the city of Orange. That same year Walmart hired her as a supplier to source fruits and vegetables from smaller farmers. "Every time we had a crop, I would think about the social ramifications of that crop and the impact to society," Montoya recalls. She saw firsthand the struggles faced by growers, especially small farmers, including lack of up-to-date and relevant information. She began brainstorming ways to help.

Montoya started AgTools in 2017, enlisting both her brothers, who have engineering backgrounds, to help build the platform. The company's product was introduced the following year at the Produce Marketing Association's annual conference.

So far, AgTools has been sold to some twenty fruit and vegetable shippers and producers. The company interfaces with some seventeen industries, including transportation, banking, insurance, and government. AgTools is available in English, Spanish, and Portuguese.

In 2019 Montoya partnered with Charter Communications and the Western Growers Association in Salinas to launch a pilot program that trains female farmers to use AgTools on their farms. In October of that year, Montoya held a training session with the first such group of farmers in Gonzales, a city just south of Salinas.

Says Montoya: "The bottom line is: can they make more money with this tool?" She notes that female farmers remain a minority in the Salinas Valley and throughout California. If successful, Charter Communications may help extend the program to several hundred female farmers.

Montoya has a passion for agriculture. Photo courtesy of Amy Vong, UCI Beall Applied Innovation.

COVID-19 did not appear to stop Montoya's plans as she stepped up raising capital for the company. In March 2020 AgTools won a $250,000 award from the San Diego Angel Conference held at the University of San Diego. AgTools also launched its new slogan, "See More|Achieve More." Montoya said sales and interest were growing rapidly due to COVID-19 because growers and shippers globally, impacted by the pandemic, were demanding more and better information to help them make productive and effective decisions. In June 2020 AgTools won the Flywheel Investment Conference prize that came with a $125,000 investment award. The service, according to Montoya, is now used by some of the world's largest fruit and vegetable suppliers.

Agtech's female entrepreneurs continue to face barriers, Montoya observes, including capital, age, experience, sex, and race. "The investment world believes more in younger generations with ideas than mature, seasoned businesspeople. Add in factors of gender and race, and our opportunities for funding are much smaller," says Montoya,

Montoya shares more about her innovation. Photo courtesy of Amy Vong, UCI Beall Applied Innovation.

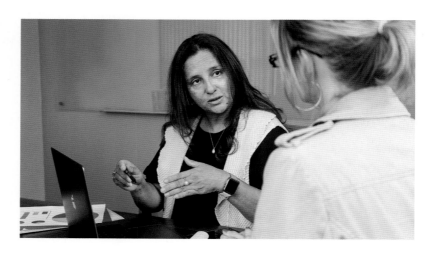

a Baby Boomer. Her key to success has been finding other women and other nontraditional investors who believed in the cause. Luckily, there are many of them.

This profile is based on an article published in a series about female innovators in agtech in Techonomy *in 2019, a digital publication that focuses on technology in society.*

Citation

Amy Wu, "This Woman Entrepreneur Gives Farmers Information—to Grow," *Techonomy*, November 27, 2019, https://techonomy.com/2019/11/this-woman-entrepreneur-gives-farmers-information-to-grow.

At a program for women farmers in the Salinas Valley in 2019, attendees were trained in how to use AgTools. Photo courtesy of Martha Montoya.

ELLIE SYMES

The Bee Corp, Indianapolis, Indiana

Software Helps Almond Growers Monitor Hive Strength

Ellie Symes is beekeeping, surprised that the hive is collecting so much honey that it is taller than she is. Photo courtesy of Ellie Symes.

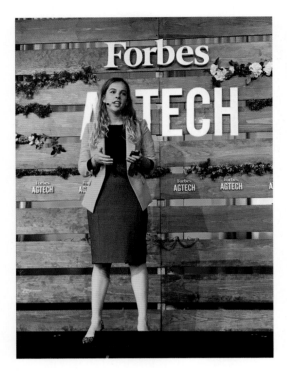

Symes at the Forbes AgTech Summit.
Photo courtesy of Ellie Symes.

Ellie Symes could easily pass for the girl next door. The petite brunette, with her dimpled ear-to-ear smile, is affable and easy to converse with. Symes is in her mid-twenties, and her youthful persona not only makes her stand out in a sector known for being male-dominated but belies her expertise: she is a ninja when it comes to talking bees and bee pollination.

Symes launched her agtech company, The Bee Corp, in 2016. The Bee Corp offers a software platform that monitors and grades the quality of hives before pollination. The technology, delivered through a mobile app, helps beekeepers and growers make sure their hives are healthy and ready to go. The start-up includes Symes and a small team of engineers and researchers who are based in Indianapolis, Indiana.

At the Forbes AgTech Summit in Indianapolis in 2018, she held the stage during the "Show Me the Honey: Innovating to Insure Healthy Bees and Honey" panel. During the question-and-answer portion before an audience made up of top executives, Symes fielded a question on what her company's technology is taking pictures of. She explained that the Bee Corp's Verifli is an infrared image analysis tool that helps growers measure pollination value. It captures an infrared image of the clusters of bees inside the hive box from a device that is attached to the user's smartphone. "It's an infrared image and we are reading thermal heat coming off the hives, produced from the body heat of the bees. The bees are actually creating body heat to heat the eggs, just like a bird sits on a nest," Symes said. Once the image is uploaded, a prediction of hive strength is made in four seconds and shared with growers.

The company's tagline is "Snap. Grade. Go about your day." "We focus on inspection and ensure growers they have strong hives and make sure they measure their pollination contracts. We can inspect earlier in pollination and be able to make decisions," Symes said, noting that the technology is such that the process does not require the hive boxes to be open, thereby avoiding manual inspection.

Symes at The Bee Corp booth at the Almond Conference in Sacramento in 2019.
Photo courtesy of Ellie Symes.

Something New

Symes was an undergraduate at Indiana University in Bloomington when she began working on the project that would later become The Bee Corp. She has no farming background and fell into both entrepreneurship and agriculture. At the end of her freshman year, as she was looking beyond the prospect of lifeguarding again in the summer, Symes began researching internship opportunities in areas of the environment and ecology: "My goal [at the time] was to build a career and get the studies I needed to solve environmental problems."

A beekeeping internship piqued her curiosity. "It was something new and different," she recalls. Little did she expect that she'd get hooked on beekeeping.

She returned to Indiana University that fall with the idea of starting a beekeeping club and program, thinking that it would be a good intersection between agriculture and technology. The club was colaunched by Symes and Wyatt Wells (the company's chief marketing officer) and included other students. The club was a success, and eventually Indiana University Foundation board members approached Symes with an offer to help take the club to another level. They had also identified

Photo courtesy of Ellie Symes.

the challenges associated with honey bee health as an area they were keen on exploring further.

"They said we love what you've done with the club, but we want you to dream bigger, and we want to help you," Symes says. "They saw a very energetic person who was very enthusiastic about honey bees. I think starting the club and beekeeping program was very entrepreneurial."

The IU Foundation encouraged Symes—now enrolled in a master's program at the university—and a small team of fellow students to enter the university's BEST Competition, which awards a $100,000 prize. The business plan competition for Indiana University students is a partnership between the School of Informatics and Computing and the Kelley School of Business.

The team's business plan, which centered around a company of tech-enabled beekeepers, took home the BEST Award. The award was a key milestone in that it allowed Symes and the team to launch a business plan and commence a round of funding. But in the face of real-world constraints the plan soon evolved into other areas. "We quickly realized that building a commercial beekeeping operation is really capital intensive and building sensors is silly," Symes recalls. "Since then we've evolved the model several times." During this time the team came up with the company name and decided to incorporate it.

Seeking Problems and Solutions

After graduating from Indiana University with a bachelor's in environmental management and later a master's degree in information systems, Symes and company took the idea of the initial product that tracks the life cycle of the hive and ran with it. Within two years of winning the BEST Competition, The Bee Corp launched Queen's Guard, its first product designed for the pollination space. It also

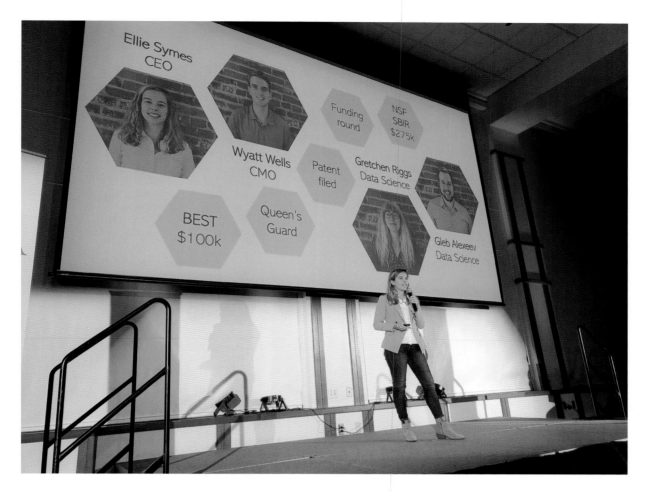

Symes speaking at the gBETA Pitch Night in Indianapolis in 2018. Photo courtesy of Ellie Symes.

received $275,000 from a National Science Foundation Small Business Innovation Research grant to solve pollination problems, including the cost of pollination and quality of the beehives.

The grant allowed Symes to retain a team and also engage in a discovery phase. They set out to learn about some of the biggest challenges that almond growers faced, and they ended up conducting over one hundred interviews. One clear problem that growers face, for example, is the rising cost of pollination, which now even surpasses the cost of irrigation and harvest.

The information garnered was golden and helped the team narrow its focus and create Queen's Guard, a product that used temperature sensors to determine whether the queen bee (which lays the eggs) was alive in a beehive or not. The team also launched a product named QGPS that uses encrypted GPS data to alert beekeepers and local authorities about stolen hives.

One of the FLIR camera models used in Verifli. Photo courtesy of Ellie Symes.

Pollination is vital to agriculture, and honeybees play an essential role as crop pollinators. Symes zeroed in on California almonds, which was a wise choice considering the breadth of that large market and the essential role pollination plays for almond growers.

California is by far the largest almond producer in the world, producing 80% of the world's almonds and 100% of the US commercial output. California's Central Valley, with a climate that is favorable for almond growing, is the unofficial mecca of almond production. Statewide the crop is valued at nearly $6 billion annually, according to the Almond Board of California.

Honeybees and almonds are heavily dependent on each other as almond trees offer the bees a food source every spring. At the same time, almond trees are not easily wind pollinated and most varieties require cross-pollination (the transfer of one tree variety to another). California almond grower Stuart Woolf uses The Bee Corp's software and says such technology helps him and other growers greatly as they often rent hives from beekeepers out of state. Hives are shipped in from all over the country for almond pollination, since there aren't enough hives in California to do the job. This journey is often stressful for the hives, so it is important to check them upon arrival at California almond orchards to make sure they are strong enough to pollinate effectively.

Members of the fledgling Bee Corp team faced many hurdles: the company was a startup, it was not based in California, where most of the almond growers are, and it had a tough audience (farmers are some of the busiest people in the world). From their coworking space in Bloomington, Indiana, team members cold-called almond growers in California to establish relationships.

"Once they are on a phone, it's not a problem. Honestly, it's getting them on the phone [that is

Symes with her mom, Anne Symes, at a cabin in Brown County, Indiana, celebrating Ellie's master's graduation. Photo courtesy of Ellie Symes.

challenging]. You just have to catch them when they are driving in a truck," Symes says, laughing. "We reached out just trying to learn about them. It's a lot of time on the computer to find lists of people you can reach out to; it's a lot of grit and follow-up."

In February 2019 the company launched the above-mentioned Verifli.

Another company milestone was getting accepted into THRIVE, a competitive agtech accelerator run by SVG Ventures out of Silicon Valley. Symes credits the accelerator for helping connect The Bee Corp with major agribusiness players. Since then the company has been constantly tweaking its technology to meet the needs of growers. It has added filters for data sorting and a daytime value to account for the change in weather throughout the day. The technology helps beekeepers price their beehives and increase pollination effectiveness. The company considers manual inspectors its main competitor.

In September 2019 The Bee Corp won a second grant from the National Science Foundation in the amount of $750,000, which went toward boosting sales and developing a web app.

Looking Globally

As of April 2020 The Bee Corp had six full-time staff and a small group of interns and closed another round of funding at $1 million. This was no small feat considering funding was at that time drying up for start-ups because of the COVID-19 crisis. Part of the funding came from a $250,000 injection from Indiana University's IU Ventures group. "They [The Bee Corp] pursued beekeeping and solving problems for beekeepers out of passion," said IU Ventures associate Samantha Ginther in the news release. "Through strategic utilization of grant funding, they converted that concept into a well-researched product."

In February 2019 the company launched Verifli, the mobile app that allows growers to capture the data easily. Photo courtesy of Ellie Symes.

At the same time, the company had imaged over twenty-eight thousand beehives and completed its second pollination season with almond growers. Considering the global economic impact from the COVID-19 pandemic that gained speed that same month, Symes says it was fortunate the funding round and pollination season were complete. Almonds are also one of the least labor-intensive crops. "From a product standpoint we are unchanged," she says.

At home in Indiana, Symes has gained a high profile in the media and within the world of agbio and agtech. She continues to sit on the board of AgriNovus Indiana, the state's agbioscience economic development arm, and is focused on bolstering the agbioscience sector in Indiana by connecting the public and private sectors and entrepreneurs.

Although moving to California appears to make sense for The Bee Corp, considering its target market of almonds, Symes is committed to keeping the home base in Indiana, where she and her team enjoy a strong support network and lower business costs. They look forward to pursuing their

A bee in action. Photo courtesy of Trav Williams/Broken Banjo Photography.

goal of eventually extending their technology to crops all over the country. This is home. On the personal side, Symes notes she and her fiancé bought a house and are "having a lot of fun with the home projects!"

"We are thinking long-term and what is best for the business as a whole. While we are laser focused on almonds in California, we are working towards our next crop market," Symes says matter-of-factly. "We are looking globally." The sky is the limit for this high-spirited founder.

Citation

Honey Bee Conservancy, "Indiana University's Keepers of the Bees," October 10, 2015, https://thehoneybeeconservancy.org/2015/10/10/indiana-university-beekeepers.

Symes touring almond orchards with almond grower Dennis Soares of RPAC LLC in 2018 in Los Baños, California. Photo courtesy of The Bee Corp.

CHRISTINE SU

PastureMap, San Francisco, California

A Curiosity About Foods Leads to an Unlikely Path of Creating Software for Ranchers

Christine Su. Photo courtesy of Christine Su.

"Self-care is not an act of self-indulgence, it is self-preservation, and that is an act of political warfare."

—Audre Lorde

Christine Su, a second-generation Taiwanese American, dreams of dumplings, all flavors and textures—mapo tofu, Peking duck, and the traditional pork and chives. Su's latest enterprise is Kinship Foods, an online-based business that sells dumplings that celebrate the connection between food and cultural identity. The company makes it a point to source ingredients from Asian-food growers, including Leslie Wiser, founder of Radical Family Farms in Northern California who specializes in growing Asian vegetables. According to Su, the meat for the filling comes from pasture-raised pigs. Since launching the business in late 2019, the dumplings have been available for delivery in the Bay Area.

The idea was ignited from Su's desire to see Asian Americans represented in the farm-to-table movement and "for ethnic foods to also be seen as clean, organic, and regenerative." Moreover, Su observes how dumplings connect family and friends. "Dumplings are something you make together and there is this idea of togetherness and community," she says.

Su loves making dumplings, In 2019 she launched Kinship Foods, which focuses on Asian comfort food. Photo courtesy of Chava Oropesa.

Kinship Food is Su's second start-up after her flagship company, San Francisco-based PastureMap, was launched in 2016. PastureMap makes a subscription-based ranch management software tool that allows ranchers to manage their holdings more efficiently by tracking information such as cattle, soil, rainfall, and quality of grasslands. Ranchers can monitor their cattle and ranches through mobile platforms, including tablets and smartphones. The startup has an overall mission of building the regenerative beef industry (regenerative agriculture is understood as farming and

Su and a friendly calf at TomKat Ranch, an educational ranch in Pescadero, California. TomKat Ranch was an early incubator of PastureMap's prototypes. Photo provided by Christine Su.

grazing practices that help increase biodiversity and enhance the top soil). It aims to use the grazing records, soil data, and rainfall data attained through its product to help ranchers make more profitable decisions while building healthy grasslands. Su says the impetus for PastureMap began with her own fascination and curiosity about food allergies.

Although seemingly different on the surface, the start-ups share a common theme in that both are driven by missions that Su feels strongly about: supporting regenerative agriculture and celebrating cultural heritage through food.

An Unexpected Path

Su never imagined she would be an entrepreneur, but her own personal connections with food led her into the start-up world. While attending Stanford University, she often found herself covered in hives after eating. Tests indicated that a handful of allergens, including dairy and red meat, were causing her body to react. This condition pushed her to find a solution.

The native Californian went on a full elimination diet, working to source and cook all of her food from scratch. She combed through farmers markets, specialty stores, and even cattle ranches to find "food that wouldn't make me break out in hives." She found that meat and cheese bought directly from farmers didn't cause outbreaks as grocery store food items did. This discovery led to conversations with farmers about the differences in their products. Although she graduated in 2008 with a degree in political science and moved to Hong Kong to be closer to her parents, food safety and production remained passionate topics. In fact her interest grew during her travels throughout Asia, where problems with food safety and production remain prevalent.

She returned to California in 2012 because she wanted "to make positive contributions in the food system," enrolling in Stanford to earn her MBA and a joint master's degree in land use and agriculture. On top of these commitments she made time to work on farms on four continents and

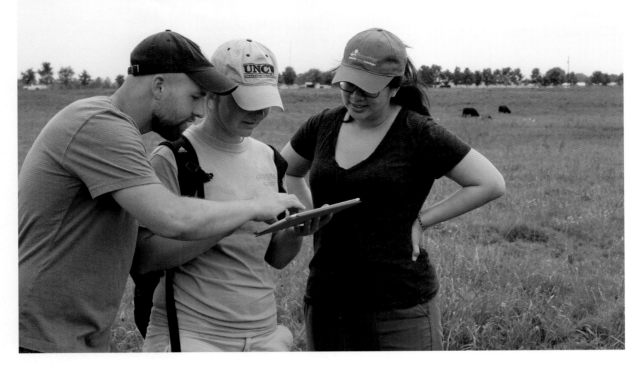

Su helps Jake Tommerdahl and his sister keep grazing records in the field. The Tommerdahl siblings are taking over and using regenerative practices on their family farm in Greensboro, NC. Photo courtesy of Christine Su.

attend agriculture conferences. She gained insight into agriculture by working at ranches in the Bay Area, including TomKat, a grass-fed cattle ranch based in nearby Pescadero, San Mateo County.

Although Su didn't know where she wanted to work in the food industry, she says her passion for agriculture was certain: "I think farmers are owed a great deal of gratitude from city people like me who don't really think about where our food comes from." In 2014, with her MBA and several internships under her belt, she colaunched PastureMap with George Lee, who was the company's chief technology officer. Their goal was to develop a tool that could solve common problems that many ranchers faced. "I kept seeing these huge paper charts and maps stapled to the barn wall, and stacks of notebooks, and realized that ranchers run their entire business on pen and paper," says Su.

Lee and Su and their small team began developing the software that is the core of the product, and Su began attending industry conferences to introduce the product to ranchers.

Many ranchers have praised PastureMap and said the platform has saved them money. Carrie Richards is a fifth-generation cattle rancher in California and part of a family of ranchers called Richards Grassfed Beef, which produces regenerative beef in Oregon House, California. Over the years, Richards and her siblings noticed consumers were demanding more and better-quality information about their beef. In addition, the family wrestled with tracking the growing amount of data their operation generated, data that is required for certification by the American Grassfed Association.

Su leading the Next Generation Producer mainstage panel at at the 2019 Grassfed Exchange in Santa Rosa, where she was the conference cochair. It was the first mainstage panel in the conference's history that included a majority of women and farmers from the LGBTQ+ community. Everyone was under forty. Photo courtesy of Christine Su.

They have found that using PastureMap makes data collection much easier. Richards and her team used the app to view the property, zoom in from where they are standing, and take before-and-after photos of everything from plant species and livestock. They are also able to track soil and vegetable points. "You can measure each pasture," she says, noting the data can then be easily sent out for analysis.

Their annual hay bill down-spiraled from $130,000 to $90,000 as better management helped them identify areas that required fencing. Richards credits this and other savings to holistic management and PastureMap's mapping tool (as opposed to eyeballing the land), noting that the product helped them narrow their acres to a more precise size and ultimately saved them costs.

An Anomaly

Su is somewhat of an anomaly in the agtech sector. She is relatively young, the daughter of Taiwanese immigrants, and a women, and she doesn't come from a long line of farmers.

At industry gatherings, Su observed that "everyone on stage looks a certain way, and it's not like me." Given the predominantly older, white, and male makeup of ranchers, "I am often the only minority and young woman under forty at a conference of over a thousand," she says. In the early days she wore western boots and trucker caps to try to fit into the mix, but she found that instead of drawing positive attention to her product, they made her look more like a novelty. Given this, she has a passion for diversifying the workplace. She cites Google's annual diversity report, which shows progress in hiring women, but little change with offering positions to people of color. Su strongly believes that differences in culture, gender, class, education, race, and ethnicity offer a "diversity of perspectives."

Although PastureMap's workforce is majority women and people of color, Su acknowledged there's room for improvement. "We look at the diversity metrics and think how we want to do better. Right now, all developers here are male." To help in that directive, she tapped Code2040, a nonprofit that promotes black and Latino engineers to recruit developers who are people of color. Su also scours agricultural graduate programs with a focus on ranching and stays closely connected with

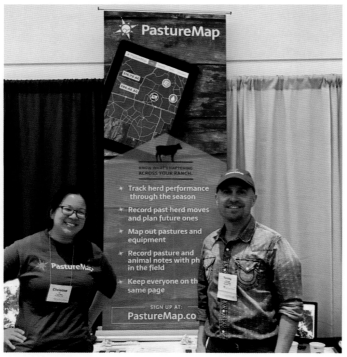

Su and PastureMap teammate Tommy Latousek at the Land Trust Alliance in Denver, Colorado, in 2018. Land trusts like The Nature Conservancy use PastureMap to help their ranchers achieve conservation goals. Photo courtesy of Christine Su.

her alma mater, Stanford University. Several PastureMap staffers are Stanford grads.

PastureMap has been on a growth trajectory; as of late 2019 it had six employees working out of San Mateo and San Francisco. At one point it was based at the now-defunct Impact Hub, a spacious and shared workplace in the heart of the city's Mission District that houses socially conscious start-ups.

Since 2017 PastureMap has raised $3.5 million, including $2.3 million in a seed round led by Eric Chen of Ovo Fund and $1 million from friends, family, and grants. Another early funder was Sallie Calhoun, the owner of Paicines Ranch in San Benito County. Calhoun is a former Silicon Valley engineer who is a staunch proponent of regenerative agriculture. Calhoun first met the young founders of PastureMap when they were at TomKat. Calhoun said that she was skeptical the company would make money but that the founders were worth investing in. "Christine is a wonderful force to have in agriculture," says Calhoun.

Today PastueMap has more than ten thousand users in forty countries.

"We are very excited about the international expansion," says Su. Over the past year, she has traveled to major industry conferences in the Midwest and has spoken at a dozen agtech and industry conferences, including the Grassfed Exchange, the BeefTech Conference,

PastureMap's mapping technology allows ranchers to track their herds and daily grazing records and figure out where and when to move animals to new pastures—all from their phones. Image courtesy of Christine Su.

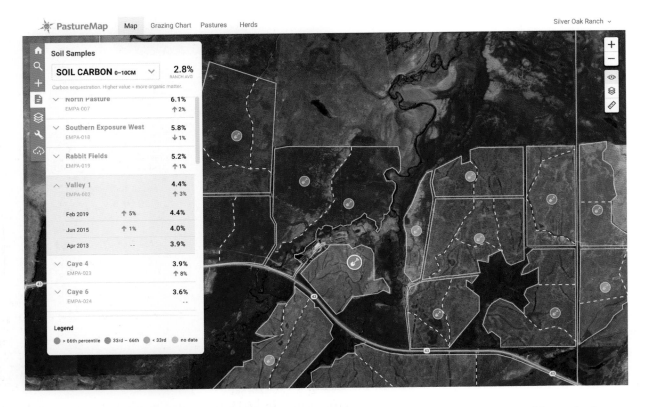

In 2019 PastureMap launched the first USDA-funded public-private partnership to help ranchers upload soil sampling data and measure how grazing practices could improve soil health on a mapping platform. Image courtesy of Christine Su.

and the Skoll World Forum. PastureMap continues to seek talent in engineering and development in line with Su's steadfast belief that diversity and inclusion are critical to the bottom line. She is cautiously optimistic that change is happening. Because minorities are already the majority in California, Su calls inclusiveness an "inevitable shift."

Diversity and Inclusivity

In the meantime, she takes pride in knowing that she is working to promote diversity and offer opportunities to women and people of color who share her passion for cattle ranching and food production. "Representation just by being who we are is already an important message," she says with a smile. "I hope that we are able to cultivate the next generation of minority female leaders in ag and in tech."

Su enjoys the dumplings she and her cousin Joyce Lin made. Photo courtesy of Christine Su.

Su continues to step up her efforts in increasing diversity, especially when it comes to uplifting the voices of Asians and Asian Americans in agriculture. In 2019 she began pulling together a network of names and contacts of Asians in agriculture with similar interest in celebrating food, regenerative farming practices, and cultural heritage and loosely called it RegenerAsians. She proposed and ultimately led a discussion group for Asian farmers at EcoFarm, a premier conference in organic farming. In January 2020, at the annual EcoFarm conference in Pacific Grove, California, dozens of Asians working in agriculture attended the discussion group and shared the challenges they faced and also the opportunities they foresaw.

Su operates in the world of agtech start-ups, and she says that change is inevitable, although it can be at a snail's pace. Su's energy, passion, and dimpled smile go a long way in contributing to change. In 2019 she cochaired The Grassfed Exchange in Santa Rosa with Maggie Gentert, a conference that drew nearly six hundred grass-fed livestock farmers and ranchers from all over the country. It was the first time in its eleven-year history that the conference cochairs were women. The chairs achieved their goal of having over 50% of the speakers be women and people of color on stage. A third of the mainstage speakers were indigenous farmers from various countries talking about regenerative agriculture in their context.

Representation is critical as "it fixes the distorted narrative that farming is a predominantly 'white' space. Globally, the majority of food is produced by smallholder women farmers," Su says. "I hope the message is clear: don't just include us. Put us in charge."

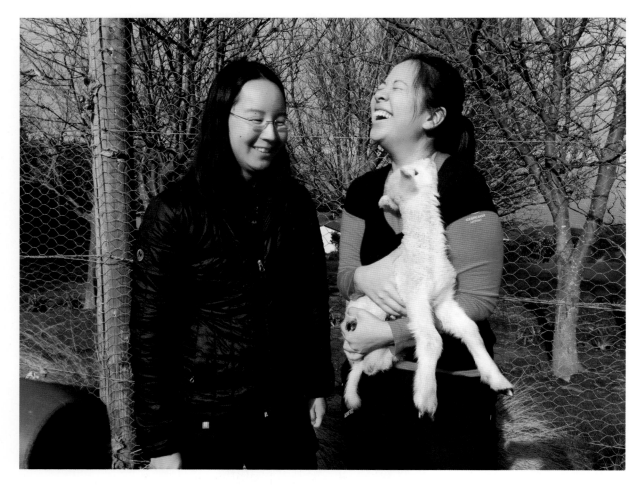

Su and Jennifer Tsau, the company's first chief technology officer and an early cofounder. Christine and Jennifer spent a summer in New Zealand as entrepreneurs-in-residence at the New Zealand Merino Company, doing pasture management research and learning from the best sheep- and cattle-grazing managers in the country. Photo courtesy of Christine Su.

Going into 2020

In April 2020 Su sold PastureMap to Soilworks Natural Capital, a Texas-based public benefits corporation with a focus on regenerative agriculture that she says will continue to invest in the company "so I can take a step back." She joined Food System 6 Accelerator, which provides strategy and executive coaching for CEOs of food businesses in Redwood City, California, as an entrepreneur-in-residence. Kinship Foods was put on hold due to the COVID-19 pandemic.

As part of building her business, Christine regularly connects with ranchers. Photo courtesy of Christine Su.

Citation

Amy Wu, "Challenges and Opportunities: Christine Su—CEO and Founder of PastureMaps," February 16, 2017, https://www.thecalifornian.com/story/news/2017/02/16/challenges-and-opportunities-christine-su-ceo-founder-pasturemap/97983862.

MARIANA VASCONCELOS

Agrosmart, São Paulo, Brazil

A Daughter of Farmers Creates Smart Software to Boost Crop Production

Mariana Vasconcelos speaking at the World Economic Forum in Davos, Switzerland, 2020. Photo courtesy of Mariana Vasconcelos.

Photo courtesy of Mariana Vasconcelos.

"Don't ask what the world needs. Ask what makes you come alive, and go do it. Because what the world needs is people who have come alive."

—Howard Thurman

In early spring of 2020 Mariana Vasconcelos found herself at a crossroads. With the COVID-19 pandemic spreading globally, the CEO of agtech start-up Agrosmart had to decide whether to cut costs, including staff. The timing seemed uncanny, if not potentially devastating, for in the growing landscape of start-ups, Agrosmart is a rising star.

Mariana Vasconcelos launched Agrosmart in 2014 as cofounder and CEO.

Agrosmart monitors crops and provides farmers and the entire food supply chain with agronomic and traceability insights. Headquartered in Brazil, the company uses data acquired from sensors installed on farms to monitor crops, with the goal of boosting crop production. The company's mission is to "make agriculture more productive, sustainable, and resilient to climate change."

Vasconcelos grew up in a farming family and says that she is driven by wanting to help all farmers be more efficient and increase their yield.

The company has built itself up to nearly sixty staff and to date has raised $8.8 million. Vasconcelos, despite her youthful façade, is a dynamic presence and an engaging and passionate speaker. In 2020 she was included in *Worth* magazine's list of Groundbreaking Women of the Year.

But the pandemic has threatened to stop the company's growth momentum and has created an additional layer of challenges for farmers globally. With schools and institutions closed and the food systems derailed, farmers, especially small-size producers, needed to find new distribution channels for their products, whether it be through launching food coops or online business platforms. For Vasconcelos, COVID-19's impact on farmers hit home in a personal way. She is the daughter of farmers and when younger witnessed the problems they faced. The family farm in Pedralva, Brazil, centered on sugarcane (Brazil is the world's largest exporter of sugarcane) and in recent years shifted to corn and horse breeding; her brother continues to grow organic vegetables and coffee. In this challenging profession, the fragile line between success and failure during a growing season was heavily dependent on Mother Nature and a grower's intuition.

Photo courtesy of Mariana Vasconcelos.

Photo courtesy of Mariana Vasconcelos.

"I suffered seeing my Dad not knowing if what he was doing was right or wrong, trusting his knowledge and [what] passed through generations," said Vasconcelos in a 2017 speech at a precision agtech industry conference.[1]

She said her father looked to neighbors to make decisions. "He said, 'If they spray, then I spray. If they harvest, then I harvest' . . . A lot of farmers depend on intuition, not knowing exactly what to do," she has stated in interviews.[2]

Vasconcelo's personal history inspired her to launch Agrosmart.

She is a staunch believer that innovation, specifically digital agriculture, is an essential tool in helping farmers—especially small- and midsize producers. It was the growing problems related to climate change "that made me want to bring technology to farmers so they could be more resilient," she says.

With that in mind, Vasconcelos made the decision to maintain as many staff as possible. She also cut the price of her company's products to make them more affordable for small farmers. "We are keeping a full team to support the farmers to keep producing, no matter what, and we have reduced our prices so farmers can access technology," she says.

Even during the pandemic, interest and demand for Agrosmart have seen an uptick as farmers turn to technology as a solution to cutting costs and increasing yield.

Vasconcelos observes: "Farmers are looking for more technology and need more technology in order to be able to understand what has been going on in the field, especially with the lack of labor. . . . They have to keep workers very focused and isolated [social distancing] as much as possible. We also have very limited imports and transportation is very limited."

Merging Passions

Despite growing up on a farm, Vasconcelos had little interest in a career in agriculture, much less returning to or running the family farm,

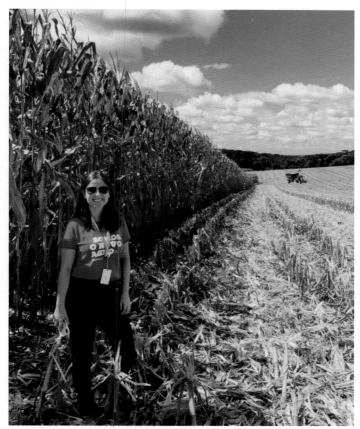

Vasconcelos in the cornfields in Ponta Grossa, a city in southern Brazil. Photo courtesy of Mariana Vasconcelos.

which her father started before she was born (her mother, now retired, was a schoolteacher).

"I was not very interested in farming and I really didn't want to stay on the farm," she says. In a 2017 *Financial Times* article she admitted, "I thought it was very boring. I liked the horses, but I did not like the farming."

When Vasconcelos was 16 her father handed her the keys to the family bakery, which her grandfather started, hopeful that she'd learn how to run it in case he passed away at an early age as his father had. The bakery was in many ways a companion to the family farm, inspired by a farm-to-table vision of offering coffee and fresh foods. She would continue to run the bakery until Agrosmart took off.

After high school she studied business at the Federal University of Itajubá, a region about 270 miles outside São Paulo known as Brazil's Silicon Valley. She concentrated her studies on tech companies and start-ups and quickly became interested in an area called the internet of things. Popularly

Vasconcelos, third from left, at the Forbes AgTech Summit in Salinas in 2019, speaks at a panel moderated by John Hartnett, the head of SVG Ventures, on the far right. Photo courtesy of SVG Ventures.

known as IoT, the internet of things refers to a system of internet-connected devices, including sensors that transfer data over a wireless network. She wondered how IoT could play a significant part in improving efficient in a variety of sectors.

"I had a lot of contact with hardware, sensors, and robotics," says Vasconcelos, recalling the frequent tech-related events at her university, including robot war competitions in which business and engineering students were often matched.

After finishing college, she moved to Germany to enter the training program at Robert Bosch GmbH, where she worked in sales within the automotive industry. While at Bosch, she followed the development of IoT technologies to search for a way to use innovation to help a variety of industries.

Upon her return to Brazil she wasted little time in launching her first start-up. She tapped a student from her college days whom she had participated in robot competitions with and in 2012 cofounded SmartApps, a consulting company with a focus on using IoT solutions for a wide range of sectors, from oil to medical to environmental industries. Quickly realizing that the company as constructed was not scalable, they shifted their focus to developing concepts for a scalable company. Mariana and her team created an open source platform where they could connect with other innovators and where developers and engineers could create their own solutions (such as building sensors). They also organized hackathons to incentivize developers to create solutions.

Vasconcelos founded Agrosmart when she was twenty-three. Photo courtesy of Mariana Vasconcelos.

After a while, "We saw the connections between [the] environment and agriculture, [which] are very complicated. . . . We thought, Why not do it for agriculture?" she recalls. This led to her testing the SmartApps platform on her father's farm. "We used it on my dad's farm and we realized the interest was very high so we decided to launch the company," she says, referring to Agrosmart.

Her realization of the powerful connection between technology and agriculture led to her decision to leave SmartApps and launch a tech company that solely focused on the agriculture market. A year later, in 2014, Vasconcelos, then twenty-three, and two childhood friends, Raphael Garcez Pizzi and Thales Nicoleti, founded Agrosmart, which creates software that uses artificial intelligence to make agricultural predictions based on data from the soil, weather conditions, and the genetic properties of the crop. The software, which can be accessed on mobile platforms, analyzes the raw data collected from field sensors (which is transmitted via satellite or a lower-broadband network) to make recommendations on everything from seed placement to climate modeling.

The software uses quantitative methods to simulate the interactions of the important drivers of climate and other factors to forecast what lies ahead. There are a variety of products on the platform that address irrigation management, weather forecasting, the traceability and sustainability of food companies, and financial intelligence for banks and insurance companies.

"Agrosmart is a platform that is generating data on the ground and in the farm and delivering intelligence of that data across [the] food supply chain, making it more productive and more sustainable," she says.

Vasconcelos notes she and her cofounders had an edge in that all three have parents who are growers, and they understood the challenges in farming. They shared a common belief that data

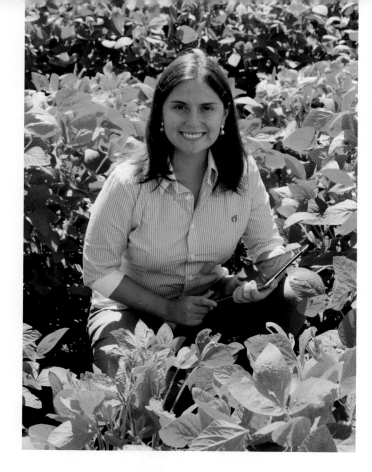

Vasconcelos comes from a family of farmers. Photo courtesy of Mariana Vasconcelos.

science was the solution to the roadblocks in food production.

The start-up was fortunate in that it had a ready-made base of operations— the bakery that Vasconcelos continued to run. The team also had a small base of farmers, many of them friends with their parents who had watched them grow up. The growers were eager to help and test the technology and offer feedback, but they were slow to make a commitment when it came to purchasing the product. Vasconcelos believes that age and gender were added barriers to sales and fundraising.

"There were many times that agronomists did not put too much faith in what I was doing; the farmers themselves were very open to receive me and they treated me like a daughter, but they didn't really want to pay me or do business with me," she explains. The solution? "We built a more diverse team with men and more senior people for sales," she says. So far, the mix has worked and led to a rise in sales and investment.

Agrosmart asserts that its innovation helps farmers increase crop yield and save water. The majority of its customers are in Latin America (many of them in Brazil), although the platform is available globally. Since the platform was launched, customers have saved up to 40% on energy and input costs and 60% on water usage, and they have upped their productivity by an average of 20%. Vasconcelos says the company now monitors more than 3.5 million acres of farmland and works with major corporations, including Coca-Cola, which uses the technology to monitor fruit farms in Espírito Santo, a state in southeastern Brazil.

What has also helped the company's steady rise in sales and funding were innovation accelerators. Early on, Agrosmart participated in Google's Accelerator program, where the company received the support of engineers and business strategy experts. She also readily found mentors and support in Brazil's burgeoning agtech ecosystem, ncluding through the Baita Accelerator, which she joined. In 2017 the company was accepted into the THRIVE Accelerator run by SVG Ventures in

Silicon Valley. In addition to being a member of the THRIVE cohort, Agrosmart was recognized with an award for "Operational Excellence" from THRIVE. By 2019 the company was operating in nine countries and had more than thirty thousand devices connected to the platform. That same year it raised $5.8 million—considered a robust sum in the agtech space—led by Bradesco's Inovabra Ventures fund. Bradesco is Latin America's second-largest private bank.

Vasconcelos comes across as someone who is outspoken if not adamant about making a difference in food systems. There's a bit of advocacy interwoven into the company's DNA, and Agrosmart definitely has a voice when it comes to tackling major issues such as sustainability and the environment. She joined Thought for Food, a nonprofit that bills itself as "the world's next-gen innovation engine for food and agriculture," as a global ambassador. Since 2019 the company has been a member of the United Nations Global Compact, a collective of business leaders who uphold UN goals on sustainability.

"There are many other members of that [UN Global Compact] such as Fortune 500

Vasconcelos speaks at the Hello Tomorrow Global Summit in Paris in 2016. Photo courtesy of Mariana Vasconcelos.

companies," says Vasconcelos. "I think there are very few start-ups that act on that level. We are also doing a business but trying to play a very important role on the decision making and understanding of sustainable development and environmental goals."

Vasconcelos herself has gained the limelight in recent years by providing a mix of innovation and novelty as a young entrepreneur in a space dominated by men. In 2018 she made the "Innovators Under 35 Latin America 2018" list of leading young innovators compiled by *MIT Technology Review*. She's been ranked as part of "Forbes Under 30" and was on Fast Company's "Most Creative" list.

The number of women decision-makers in farming (women producers) was very low when Vasconcelos was a child, but she's seeing more women emerging in the space. "In agtech it seems low and there is still a long way to go, but it's growing," she says, hopeful.

She's optimistic that the agtech sector itself will draw younger people to consider careers in agriculture and will slowly dispel a stigma that farming is mostly tractors and overalls. "That plays a very important role on who wants to stay with farming. Agtech bridges those two worlds and incentivizes many of the kids to work in farming," she says.

Looking Ahead

Before COVID-19 hit, her focus was building out a corporate team, putting together a merger and acquisition strategy, and hiring representatives in continents where Agrosmart wanted to grow its presence, namely Southeast Asia and Africa.

"Our goal is to become the leader of agriculture platforms in Latin America, and the advantage is we know how to handle the various spectrums of Latin America's farmers, including those with different education levels," says Vasconcelos, who indicates her primary focus will remain Latin America.

Despite the pandemic, Vasconcelos is still fully committed to expanding Agrosmart but realizes growth might just take longer. COVID-19 and its impact on the global economy have reminded her that despite all that the team has accomplished, Agrosmart is a start-up. In April 2020 she said the team was making changes to its business strategy with cost reduction and cash preservation in mind "in order to extend the runway of the company."

"We believe fundraising will take longer," she says matter-of-factly.

The drive to transform Agrosmart into an entity that makes a lasting impact comes from the image of the small farmer. In a 2017 presentation she focused on a farmer named Mr. José. "He didn't know how much water he was using," she told the audience. "Besides his employees had to go to fields daily to understand about pests and what's going on, and their method was based on intuition. Now he's connected to his crops and he has access to [them] wherever he is. He can see how his farm [is] doing, and he can understand the relationship between environmental conditions and his results." This led to a 5% growth in the farm's productivity, a 30% savings in water used, a 20% reduction in energy expenditures, and a 128 hours cut in labor. "I think this is valuable economic value."

Back at home in Pedralva (a municipality in Brazil's southeast region), the family farm continues to run. And a milestone that she doesn't readily talk about unless reminded is that her parents have made her technology a part of their operations. They celebrate her as an innovator. Her moniker is now "CEO at Agrosmart, and daughter of farmers who believe in a connected world."

Despite the pandemic, Vasconcelos is still fully committed to expanding Agrosmart. Photo courtesy of Mariana Vasconcelos.

Since 2019 she's also been working toward her MBA with a focus on agribusiness at the University of São Paulo's agriculture college. Running Agrosmart consumes most of her day, but she does find time for other activities, such as spending time with her family and riding horses, one of her greatest passions.

Citations

Yasmin Gagne, "Her AI-driven Software Helps Farmers Save Water and Make Better Decisions," *Fast Company*, May 22, 2019, https://www.fastcompany.com/90345198/most-creative-people-2019-agrosmart-mariana-vasconcelos.

Micki Wagner, "How COVID Is Pushing Innovations in Agriculture Forward," *Worth*, July 2020, https://www.worth.com/how-covid-is-pushing-innovations-agriculture-forward.

Amy Wu, "The Female Agtech Innovations Aiding Farmers amidst COVID-19," *Worth*, May 2020, https://worth.com/female-agtech-innovators-driving-food-system-aiding-farmers-amidst-covid-19.

Notes

1. *Precision agtech* refers to the use of satellite or site-specific crop management for growing crops and raising livestock.

2. "Mariana Vasconcelos: Precision Agriculture," https://www.youtube.com/watch?v=e9SSb4ED1nw.

PONSI TRIVISVAVET

Inari, Cambridge, Massachusetts

A Passion for Innovation and Creating Grower-Friendly Seeds

Ponsi Trivisvavet inside Inari's laboratory. Photo courtesy of Inari.

Trivisvavet is CEO of Inari. Photo courtesy of Inari.

"Becoming is better than being."

—Carol S. Dweck, psychologist

The aha moment surfaced in 2012 when Ponsi Trivisvavet was at Syngenta, one of the largest agrochemical and seed companies in the world, where as the head of the Southeast Asia section she frequently traveled to connect with her clients. On one of these trips she traveled to a rural region in Indonesia and visited a series of farms that grew corn, a major staple in the region.

"I was in cornfields and this gentleman, maybe over seventy-five, came and sat next to me and said 'Thank you for bringing in the hybrid corn.' It turned out that he had seen a tremendous uptick in harvest from two tons per hectare to seven tons per hectare, and the surplus allowed him to purchase beef for his family for the first time.

"That was the turning point in my life. I saw firsthand how innovations in agriculture better the lives of smallholder farmers and their families and I decided to dedicate my career for this greater purpose," Trivisvavet recalls. For many years she kept the photo of the man sitting in the cornfields in her office as a reminder of why she continues to forge ahead in the area of seed innovation.

After leaving Syngenta in 2016, Trivisvavet became chief operating officer at Indigo, an agtech company that creates technologies to increase crop yield. Soon after, Trivisvavet was approached by Mike Mack, the chairman of Inari, who offered her the CEO position at his company. Mack described Inari's technology and she was immediately captivated by the potential of it.

Inari is a plant biotechnology company that focuses on seeds and genetics. Its mission is to transform agriculture and its impact on society, climate change, and the environment. The company's current focus is on row crops, including corn, soy, and wheat, that have a vast potential to feed the world. To solve some of agriculture's biggest challenges, Inari aims to create seeds that need less water and fertilizer and are drought tolerant. Inari's breakthroughs in row crops will be transferable to other crops in the future, including specialty crops.

"Inari has pretty cool technology and is going to turn technology into a business where products do a greater good," Trivisvavet says. She accepted Mack's offer, describing it as "a five-second decision." She was attracted by Inari's combination of cutting-edge innovation and by the chance to make an impact on the food system.

Trivisvavet and her daughter Janissa in a cornfield. Photo courtesy of Ponsi Trivisvavet.

She continues: "The reason why I joined Inari was that you can do so much for the world with this particular technology. I said, 'Wow, I wish I would have seen this technology a decade ago.' Inari's technology will be used to shape a lot of things including more sustainable solutions to handle the challenges of climate change and need for more conservation, a more progressive business model that will increase growers' choices, and the power of collaborations to maximize innovations."

Ponsi's interest in using technology and innovation to make a lasting impact for the greater good goes back to her childhood. Trivisvavet's personal history is a tapestry of East and West. She was born in Thailand, and her parents' work led the family to moving and living in a variety of places, including California, Malaysia, Singapore, and Thailand. Her mother owned a hotel and her father's family business was ocean fishing. Her role model growing up was her paternal grandfather, Mr. Koay, who was the head of a village on Penang, an island 182 miles outside Kuala Lumpur. Although Penang is known today more for its tourism and beach culture, in the 1970s it had a robust fishing industry. Mr. Koay died when Trivisvavet was young, but he had instilled in her the importance of giving back to society. This conviction has inspired the decisions she has made in her professional life.

"You've heard the saying that the best way to help people is to teach them to fish, not give them the fish. Well, my grandfather lived his life that way. That's what my grandfather did his whole life.

Trivisvavet inside one of Inari's greenhouses. Photo courtesy of Inari.

He basically taught the whole village how to catch the fish and to be self-sufficient in order to feed their families. His care and concern for others was why he was my role model," she says.

As a girl, Trivisvavet remembers her father reminding her and her brother that it is critical to ask how one could help others, and "If you're lucky, why don't you pay it forward as well." She continues, "I grew up in an Asian world where you have to help others and the question was always what else are you going to do in this world?"

The images of her grandfather fishing in his village on Penang and the questions her father broached over dinner tables surfaced the day she visited the elderly farmer in Indonesia.

"That picture is still with me to this day, this picture of this man," she says. Her passion took form: "I thought, I am going to use the technologies to help farmers. It was my time with the farmers that touched me the most and will always have a place in my heart and where I drive my passion for innovation in agriculture to better life for others."

Trivisvavet's paternal grandfather, Mr. Koay, was one of her key role models. Photo courtesy of Ponsi Trivisvavet.

Medicine and Business

Trivisvavet moved back and forth between Los Angeles and Asia during her childhood due to her parents' family businesses in Asia. She attended high school in Thailand and had plans to be a doctor, but a health situation prevented her from pursuing medicine. Instead she turned to electrical engineering, fascinated by technologies and the ability to help create infrastructure. She majored in electrical engineering at Chulalongkorn University, one of the oldest and most prestigious universities in Thailand. After a brief stint as a fiber optics engineer, she realized that engineering wasn't for her and that she enjoyed interacting with large groups of people. Simply put, "I love dealing with people," says with a laugh.

"I wanted to create [a] more immediate impact to society at large," she says. She enrolled in business school at the SC Johnson College of Business at Cornell University in Ithaca, New York, to explore a new career in business. Little did she know that the MBA was the start of a career in agtech.

After business school she was hired by McKinsey & Company, the global consulting firm, and found herself with a multitude of opportunities to work in agriculture, including with large agricultural input companies that focus on products such as seeds, fertilizers, and food additives.

"As soon as I touched on those, that's how I fell in love with agriculture," Trivisvavet says.

Three years later she jumped at a chance to work at Syngenta and move to Basel, Switzerland, where the company is headquartered. For the next eight years she took on a variety of leadership roles, from the head of seeds strategy to head of global corn to general manager of trials to leading the Association of Southeast Asian Nations (ASEAN) team. In 2014 she was promoted to president of Syngenta Seeds North America.

Inari's technology focuses on the seeds of classic row crops such as corn. Photo courtesy of Inari.

In 2015 and 2016 the world of seeds had been condensed through large deals, spurred by market forces that encouraged consolidation and cost cutting. Although new innovations were cropping up, major players such as Monsanto, Bayer, Dow, and DuPont had merged. In February 2016 Syngenta was acquired by ChemChina, a Chinese state-owned enterprise, for $43 billion after a buyout from Monsanto failed. The industry was changing due to consolidation, and Trivisvavet increasingly missed the research, discovery, and innovation. She left Syngenta.

"I wanted to pursue a different world," she says, referring to her return to innovation.

In an interview with *Authority Magazine* in 2018, she said that a central question that motivates her drive to continue in the area of seed development is "Why isn't anyone applying what's being done in the biomedical area to plant and crop sciences?" So, when the opportunity surfaced, Inari felt like fate, a key attraction being that it was "purely technology."

Seeds of the Future

Inari is a marriage of science and farming, using biology, computational agronomy, data science, and software engineering to create seeds "customized" for specific crops and environment, that can withstand changing soil health, weather, drought, and new diseases.

Inari uses its technology to edit the seeds of classic row crops such as corn so that they are grower friendly. Photo courtesy of Inari.

As of May 2020 Inari had 145 full-time staff, including CEO Trivisvavet, and it had raised $145 million. Although most of the research and development is based in Inari's headquarters, located in Cambridge, Massachusetts, a hotbed of some of the world's most prestigious academic institutions, the start-up's product development and commercial teams are in West Lafayette, Indiana—the corn belt. There its 26,000-square-foot space includes offices, a lab, a greenhouse, and product development teams. There is also a small science team based out of Ghent, Belgium.

Inari tries to capitalize on the key markets when it comes to knowledge and customer base. The company focuses on corn, soy, and wheat (classic row crops) and uses its technology to edit the seeds so they are grower friendly. The process begins with data science offering a road map for the genomes to help determine which genes to adjust. This is followed by gene editing so that the seeds of each plant, vegetable, or crop take in less water and fertilizer. In the case of soybeans, giving more seeds per pod. As a start-up, Inari primarily works with independent seed companies in the Midwest, many of which have deep and extensive networks with growers.

Inari is currently selling computational breeding (breeding through artificial intelligence and genetic knowledge) seeds and plans to commercialize the edited seeds in the future. The computational breeding seeds are a "basis for us to further develop chassis that we will edit. This provides familiarity to the customers before the edits come," Trivisvavet says. In 2021 the company is focused on commercializing corn seeds and building up a pipeline of seeds that address environmental challenges across several crops. They also plan to expand into other countries in the future.

Embracing Womanhood

Trivisvavet says although she has never let age, gender, or race deter her from moving ahead, she's well aware that she is part of the minority when it comes to women leaders in agtech.

"I have always wanted to work in the role that can make a big impact on the world. In a male-dominated field of ranch hands and tractor drivers, I became one of the few females joining the cause of feeding the world, but also changing the dynamics of agriculture and science, particularly in leadership positions," she was quoted in *Authority Magazine*.

Ponsi consults with a team member. Photo courtesy of Inari.

At the start of her career she found age a barrier, particularly at Syngenta, where many distributors were older than she was. One year she was the only woman on the senior leadership team. "But the good news about age is you can outgrow it," she says. She never had any problems with growers and customers, including seed dealers.

Trivisvavet continues: "I think I found a sweet spot. If you pretend that you want me to [be] strong like a man, that's where you make a mistake. You actually have to build on who you are and it's OK to be a woman. I did spend a lot of time with the spouses of the growers. In the farming world, whether you like it or not, the real decision makers are the wives. Growers and seed folks care a lot about the same thing: they care a lot about family values."

When hiring at Inari, she takes diversity into careful consideration too and notes that half the staff are women.

She considers herself lucky having had mentors throughout her career such as Michael Mack, former CEO of Syngenta, and now Inari's chairman; Pam Johnson, a farmer in Iowa and former president of the National Corn Growers Association; and her husband, Jadkalvin Trivisvavet, an entrepreneur.

Running a fledgling start-up area and leading fundraising aren't for the faint of heart, but Trivisvavet says the image of her grandfather, the words of her father, and a strong desire to make a difference inspire her. She's not ashamed to admit that she's a bit of a workaholic. She takes her children (a four-year-old daughter and sixteen-year-old son) to visit the corn farms with her. And she and her husband own a working horse farm in Thailand. Her philosophy has always been "One day if you find your job is your hobby, that's when you should stop looking for a new job. Meaning that for me, I was lucky enough in that I found agriculture in my hobby." It is in line with the advice that she would give her younger self: "Continue to use your brain, but follow your heart. You can only be sustainable if you work in the area that you love."

Citation

Yitzi Weiner, "Female Disruptors: Ponsi Trivisvavet Is Shaking Up Plant Breeding Technology," *Authority Magazine*, September 12, 2018, https://medium.com/authority-magazine/female-disruptors-ponsi-trivisvavet-is-shaking-up-plant-breeding-technology-f818437194cc.

FATMA KAPLAN

Pheronym Inc., Davis, California

A Scientist-Turned-Entrepreneur Uses Pheromones to Tackle the Global Food Crisis

Fatma Kaplan at a laboratory at the UC Davis-HM.CLAUSE Life Science Innovation Center in 2019. Photo courtesy of Fatma Kaplan.

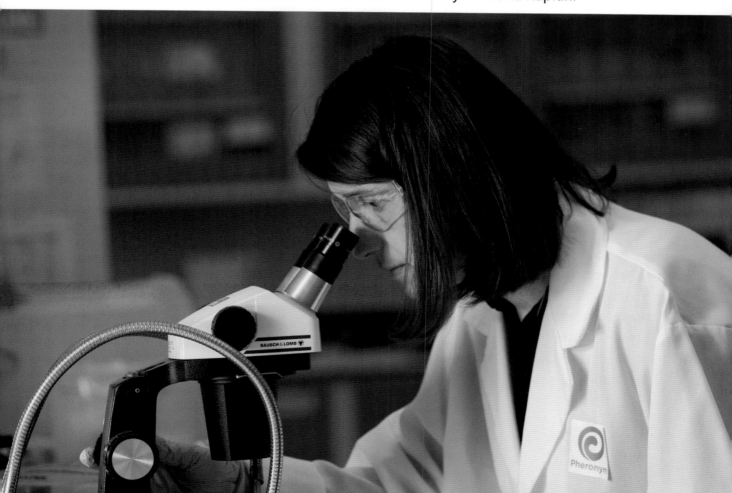

"It is better to be prepared for an opportunity and not have one than to have an opportunity and not be prepared."

—Whitney M. Young Jr., American civil rights leader

Pheronym Inc. was launched by Fatma Kaplan in 2017 with the mission to help growers control pests in an eco-friendly way. Kaplan, CEO, and her husband, Karl Cameron Schiller, cofounder and COO, run the company from its base in Davis, California. Pheronym creates biopesticides that use pheromones (chemicals capable of acting like hormones to impact the behavior of the receiving individuals) to control a wide range of agricultural pests. The pheromones that the company produces come from beneficial nematodes, also known as microscopic roundworms.

Yet Kaplan's journey from a career as an academic scientist to a business entrepreneur was unplanned, and if she were to do it again, she says she'd focus on commercializing products rather than limiting them to laboratories. Her advice to her younger self is fairly straightforward. "Skip the postdoc and start a company," she says matter-of-factly.

The daughter of specialty crop farmers in Turkey, she had planned for a career as an academic and scientist in horticulture. As Kaplan writes in a *Science* magazine column, "For years, I had been committed

Kaplan is the founder and CEO of Pheronym, a company that produces biopesticides that use pheromones to control a wide range of agricultural pests. Photo courtesy of Sarah Deragon.

Kaplan speaks at the World Agri-Tech Innovation Summit in London in 2018.
Photo courtesy of the World AgTech Summit.

to pursuing the traditional academic path because I wanted to run an independent research program and teach. I thought a tenure-track faculty position was the one job that offered both."

Kaplan was well on her way toward pursuing a career in academia. She graduated from the College of Agriculture at Cumhuriyet University with a bachelor's degree in agricultural engineering and then earned a master's degree in molecular breeding of horticultural crops at the University of Florida in Gainesville. By the time she completed her PhD in plant molecular and cellular biology and Sstress tolerance (also at Gainesville), Kaplan had obtained a significant background in biology and chemistry.

After earning her PhD Kaplan accepted a postdoctoral position at the National High Magnetic Field Laboratory, a facility at the University of Florida focused on the chemistry of purifying nematode sex pheromones and identifying their structures. At the same time she applied for tenure-tracked positions but did not receive any offers. Little did she know that the seeds of her future company were planted during this time.

Kaplan inspecting tomatoes in a greenhouse at the UC Davis-HM.CLAUSE Life Science Innovation Center in 2019. Photo courtesy of Fatma Kaplan.

As part of her postdoctoral work at the University of Florida, she purified and identified the structures of the first *Caenorhabditis elegans* sex pheromones, which she says had eluded scientists for decades. "I knew pheromones were important, but I did think someone from the agriculture industry would pick it up," she says. Kaplan was able to grow multitudes of nematodes and collect enough pheromones to illustrate their molecular structure, which answered the question as to whether *Caenorhabditis elegans* had pheromones or not and whether they used them to communicate.

Controlling insect pests with pheromones was already being done on farms, but controlling beneficial nematodes with pheromones was new territory. Kaplan had the idea that with the chemical structured identified, that knowledge could now be applied to specific plant-parasitic nematodes. In a milestone of achievement in the academic world, she published the research in the prestigious scientific journal *Nature* in 2008.

The laurel helped her land a three-year temporary position with the United States Department of Agriculture (USDA), which hired her to apply the new pheromone technology to help control parasitic nematodes (microscopic roundworms that infect plant roots). The timing was perfect since methyl bromide, the colorless, odorless toxic pesticide that had been widely used to control plant parasitic nematodes, was fast phasing out as it was known to deplete the ozone layer and can be toxic if inhaled.

"Farmers needed a solution to control plant-specific nematodes," she recalls. "For me, my heart is with farmers wherever they run their farm in the world." Agriculture is part of her personal history. As a girl she spent summers at her maternal grandparents' hazelnut farm in Düzce about 138 miles outside the cosmopolitan city of Istanbul, where she attended boarding school. Kaplan holds fond memories of the fresh produce and crisp air during her summers in Duzce.

Even though Kaplan was at the USDA, she maintained a solid plan to move forward on the academic track: she intended to attain a faculty position and continue to develop nematode pheromones. The product that her company, Pheronym, offers today—producing pheromones packaged with beneficial nematodes—had already taken shape.

Kaplan had a strong feeling she was onto something, especially when talking with fellow scientists who were directly working with farmers. "They thought this was great and would help farmers and they said yes, why not write grants and develop this technology further," she says.

Kaplan presenting at the World AgriTech Innovation Summit in London in 2018.
Photo courtesy of the World AgTech Summit.

While at the USDA she had faith that she'd find an industry partner—hopefully, a large corporation—to commercialize the new technology "because they have the resources." That said, no industry partner emerged. "We didn't get much interest. We thought with the publications [research papers for academic publications] folks might reach out to us, but it didn't really happen," Kaplan says. Time was also ticking as her appointment with the USDA was coming to an end in 2011.

She began thinking of taking the technology outside the research lab and university setting. "I thought who would be better to bring this technology to the market other than me? I had the vision, knowledge, education, passion, and the farming background."

In the meantime, her plan to become a tenured professor stalled. In 2015 she interviewed for a tenure-track faculty position, which she estimates was her 185th application in five years. "I was excited to hear that I was the top candidate. But despite that encouraging sign, once again I didn't get the offer," she wrote in *Science* magazine's blog.

Kaplan and her husband and business partner, Karl Cameron Schiller, at their laboratory. Photo courtesy of Fatma Kaplan.

Beyond the Laboratory

Kaplan attributes the search for funding and the rejection letters from her job applications to pushing her into launching what would come to be Pheronym.

"We thought we should bring it to the market, but going to investors was totally new to us," says Kaplan. She began shifting her job applications from academia to agbio and agtech accelerators and incubators. Her husband, Karl Cameron Schiller, has also been an integral part of the innovation. A trained economist, Schiller prepared the most of the budgets for grants and completed any necessary registrations, including patents.

The couple started building a team and creating collaborations with fellow scientists.

In 2015 Kaplan and Schiller established collaborations with David Shapiro of the USDA Agricultural Research Science and Ed Lewis of the University of Idaho to write grants and signed a cooperative research and development agreement with the USDA. They brought on interns in 2016 and hired Abigail Perret-Gentil as Pheronym's first employee in May 2017 (the company was formally founded in March of that year).

"We have complementary skills," she says. In 2017 Kaplan and Schiller decided to "work as if we had funding, look for investment and write grants for funding to accomplish our goals and milestones. If we did not get any funding by the end of May, we were going to close our doors," Kaplan writes on her LinkedIn profile page, referring to the laboratory under what was then Kaplan Schiller Research LLC.

At the UC Davis-HM.CLAUSE Innovation Center, Kaplan prepares NASA safety documents and certification for the AstroNematode project to send with the sample to the International Space Station. Photo courtesy of Fatma Kaplan.

Kaplan stands in the UC Davis-HM.CLAUSE Life Science Innovation's R&D field trials. Photo courtesy of the University of California, Davis.

The lucky break came when they were accepted in March 2017 into the IndieBio Accelerator based out of San Francisco, which came with much-coveted wet lab space where experiments could be conducted, $250,000 in funding, and the requirement to be based in San Francisco for four months. They moved to California, temporarily leaving behind their lab in Florida.

Coupled with funding they received from the USDA, the company—which included Kaplan, Schiller, and their team—was able to develop its first prototype and conduct proof-of-concept trials.

Their experience at IndieBio was followed with support from a litany of accelerators and incubators (some of them conducted virtually such as the Larta Institute), such as Tech Futures Group and the California Life Sciences Institute's Fellows All-Star Team (FAST) Advisory Program that focused on mentorship.

Kaplan attributes mentorship to a great part of Pheronym's success. "You can never have enough mentors. Stay tuned for 'Mentoring Pheronym,'" she wrote on her blog hosted on Medium in 2017.

It was at the California Life Sciences Institute's program where she was paired with Pam Marrone, the founder of Marrone Bio Innovations based in Davis. Kaplan and Schiller found themselves spending more time in Davis with Marrone as they developed their pitch deck (see the glossary) and product. In time, they began considering Davis as a base. They were attracted not only by Marrone's mentorship but by the city's proximity to UC Davis (one of the top schools for agriculture) and to the Salinas Valley. During this time, they also created their first prototype. "We learned how to turn a concept into something actually tangible," Kaplan says.

By the end of 2017, they decided to relocate from Florida to Davis. "Yes, I followed Pam to Davis. That mentorship was priceless," she says with a laugh. Marrone sits on the company's advisory board and has been instrumental in connecting the company with investors. She's encouraged Kaplan to make herself more visible through marketing, and the results are tangible. She's become a fixture at agtech conferences, speaking, for example, at the World AgriTech Conference in 2019.

Seeking Commercialization

Since 2019 the company has been moving forward on preliminary field trials. In 2020 the spring trials in orchards in Florida were delayed to the fall due to COVID-19. The 2020 fall trials started and will be completed in spring 2021. In the meantime, Pheronym has been able to collect data that shows the pheromones are improving the effectiveness of beneficial nematodes, even at a range of temperatures.

Some of Pheronym's milestones have been unexpected. In 2019 the company partnered with the USDA to send 120,000 beneficial nematodes to the International Space Station in Florida. That initiative gave birth to AstroNematode, an unintentional brand. The upside is that it also brought overall awareness to nematodes and the power of biological controls and publication in the prestigious peer-reviewed journal *npj Microgravity.*

"We had a couple of advisors who said it's a really cool project and it might work well with your brand," she laughs. Pheronym is the focus for now and not the AstroNematodes, although at trade shows and events there's always a great demand for the AstroNematode swag.

While not yet marketed, Pheronym's core products are biopesticides that use pheromones to control nematodes, which in turn control twenty-five different insect pests. One product is Nemastim, a bee-friendly pest control that disperses the beneficial nematodes and tackles pest insects. A second product, PheroCoat, serves as a repellent and alerts the plant-parasitic nematodes that the host plant is already infected. For Nemastim, the core customers are growers specifically in the specialty crop space, including greenhouse growers. The core customers for PheroCoat, a seed treatment, are both specialty and row crop growers.

Pheronym continues to research and develop products. It is looking to improve its existing products by increasing the effectiveness of nematodes so growers can reduce the number in the field, thus reducing the cost of the nematodes. And the company has a growing list of pests it wants to target. Kaplan and her team talked to pest control advisors and Driscoll's and learned strawberry farmers are having significant problems with the spotted wing drosophila. They connected with organic cabbage growers and Wilbur-Ellis and found that the cabbage maggot has been a significant challenge.

There been growing interest in Pheronym's products from not only seed companies but also subsegments of agriculture. Pheronym has received inquiries from cannabis companies, a poultry company, and cattle-ranching companies that are keen on controlling dung beetles and plant parasitic nematodes for organic production. Also on the upside, the company recently received funding from AgStart (an agtech accelerator) and will be hiring a technician. The company is also a finalist for the Ray of Hope Prize, which awards top biomimicry start-ups.

Back to the Farm

In many ways the journey to create and build her own company brings Kaplan back to her roots. Her grandparents on both sides were farmers and focused on specialty crops, especially tobacco. Tobacco was labor intensive and not easily profitable so they eventually switched to farming hazelnuts. Kaplan's parents divorced when she was nine months old, and she spent much of her youth in boarding school. But she also spent many memorable summers on her grandparents' farm.

Kaplan seems to have inherited her entrepreneurial side from her mother, Dilber Demir Edis. Her mother, after her divorce, juggled working for a tobacco company, managing her parents' hazelnut farm, and studying to attain her nursing certificate at the local hospital so she could be a nurse for the tobacco company. After retiring, she launched her own real estate business and expanded her hazelnut farms to fruit orchards. She is an award-winning farmer and has earned local and regional awards in the women farmer competitions held by Turkey's Ministry of Agriculture and Forestry.

Although Kaplan never considered these connections to farming and entreprenuership when she was younger, she now feels like she has come full circle.

As for the path not planned, she says her only regret is that she should have taken the entrepreneurial path earlier. "I wanted farmers to use my discoveries. I understood how to develop technology in academia, but I didn't know how to do it in practice. Now our results actually have direct impacts on farming and I've also learned how to communicate science. That is very rewarding."

Citations

Fatma Kaplan, "Making My Own Home," *Science* 354, no. 6309 (October 14, 2016): 254, https://science. sciencemag.org/content/354/6309/254.

Roger Tripathi, *Women in Agriculture*, 3rd ed., 2020, Global BioAg Linkages, https://www.bioaglinkages. com/post/women-agriculture-third-edition.

ROS HARVEY

The Yield, Sydney, Melbourne, and Hobart, Australia

Using Artificial Intelligence to Help Growers Feed the World

Ros Harvey possesses a passion for changing the world for the better through technology. Photo courtesy of The Yield.

"Seek first to understand, then to be understood."

—Stephen Covey, from
The 7 Habits of Highly Effective People

Ros Harvey never imagined that she would be the founder of an agtech start-up, much less in her prime. Soon after her fiftieth birthday in 2014 she founded The Yield in her home country of Australia.

Harvey's company uses a variety of technologies, including artificial intelligence (AI), to provide solutions to commercial growers. Her passion is to solve problems with big impact—from limited water and land supply to pest management. Harvey sees The Yield as a way to fight climate change.

In the same vein, The Yield uses technology to provide solutions to commercial growers.

In the six years since launch, the start-up has achieved key business milestones, including successfully raising money from some leading investors. As of June 2020 The Yield had twenty-seven full-time staff, including data scientists, engineers, agriculturalists, and technologists, and had raised some US$24.6 million with major investors, including Bosch, Yamaha Motor Ventures, and KPMG.

Harvey is founder and managing director of The Yield, an Australian-based agtech solutions company. Photo courtesy of The Yield.

Since its founding, both Harvey and her company have received glowing accolades from the agriculture and technology industry. In 2020 it made THRIVE's Top 50 list of scaling and visionary agtech companies. But again, Harvey says with a jovial laugh, all of this was unexpected.

She was quoted in the *Sydney Morning Herald* in 2017 as saying, "Agtech is a sexy sector. Who would have thought?" Three years later she says the sector's sex appeal has grown and more importantly matured.

Indeed at first glance Harvey appears as a wild card as an agtech founder. Prior to her current stint, she worked in nonprofits, government, and academia. Those opportunities took her as far afield as Cambodia, Geneva, and Hong Kong.

Her résumé is dizzying: she founded the Better Work program, a landmark joint program of the World Bank Group and the United Nations, and she cofounded Australia's Food Agility Cooperative Research Centre. She was also director of Better Factories Cambodia, and at one point she was deputy secretary of the Tasmanian Department of Economic Development, Tourism, and the Arts. She came with no background in farming, though her father grew up on a sheep station (farm) and has "country boy" roots.

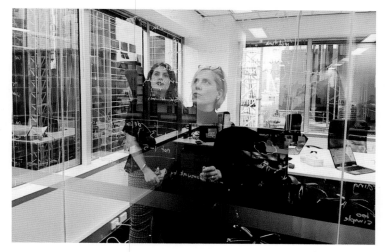

Harvey brainstorming with staff at The Yield's office. Photo courtesy of Ros Harvey.

The common theme is that Harvey's initiatives are driven by a core belief that technology is the solution to some of the world's biggest problems. At the same time there is a do-good element to what drives her business. The Yield is a for-profit business, but she hopes that the data it collects can also be used for research in supporting biodiversity/environmental sustainability.

"I have a strong commitment to creating public good with private effort through innovative uses of technology," Harvey has said publicly.

Even in running a fast-growing agtech start-up, Harvey wears a variety of hats: cofounder and strategy advisor at the Knowledge Economy Institute, cofounder and nonexecutive director of Food Agility CRC (which she cofounded with Michael Briers), and director of the Australian Conservation Foundation. Before sharing the details of her journey, Harvey issues a warning that it could feel a bit meandering.

"Amy, this could take a really long time. Have you the time for this?" she asks, along with a hearty laugh.

Daughter of a Pioneer

Perhaps she inherited her frontier spirit from her mother, Patricia.

Harvey was born and raised in Sydney, the fifth of six children in a Catholic family. Her parents were doctors, her father an orthopedic surgeon and her mother an accident and emergency doctor. But it was her mother, Patricia, from whom she might have inherited her frontier spirit. Patricia Harvey was born and raised in the outback town of Orroroo in South Australia; she moved to Sydney with her mother as a teenager, not long after her father died. She won a scholarship to attend the University of Sydney's School of Medicine and went on to become one of the first female doctors in Australia. She encouraged her four daughters "to maintain their careers whilst having a family."

Harvey says: "My mother would never describe herself as a feminist, but I think she's the ultimate feminist—it's just how it is generationally expressed." Patricia Harvey worked through age eighty-eight before she retired in 2020. She now fills her time with newfound passions such as Zumba and connecting with her twenty grandchildren.

Harvey attended a Catholic girl's school in Sydney, Australia, where students were encouraged to volunteer in their communities. "I was somewhat bored by schoolwork and more interested and focused on community volunteering and activism," she says. When she was thirteen she joined Help Mates, an organization of young people volunteering to support other young people with disabilities. Although she had goals of becoming a social worker and had a growing interest in social justice, she was also drawn to international travel and had a bit of wanderlust in her. Her first exposure to international travel was going to Indonesia with her parents.

After graduating from high school in 1981, she was awarded a Rotary International Exchange Scholarship and used it to spend a year living in Norway, as her mother's heritage is Norwegian. After she returned from Norway, she was set on working internationally.

She briefly returned to Australia and enrolled in social work at the University of Sydney, but deferred and decided to hit the road again. This

The Yield's sensing technology, here inside a poly-tunnel at Costa Group berry farm in New South Wales, Australia. Photo courtesy of The Yield.

time the destination was Tasmania, Australia's most southern state, renowned for its natural beauty and environmental campaigns. On a bushwalking (hiking and backpacking) trip she met her now ex-husband Dain Bolwell, who she says was a significant influence in her decision to work for the union movement. When she met him he was a trade union official.

The young couple married and settled in Tasmania. Harvey studied at the University of Tasmania, but her energetic personality seemed to lend itself more to work than study. She worked in the state government writing a Women's Employment and Training Strategy and also served as executive director of the Youth Affairs Council of Tasmania. The work extended to labor unions, including stints at the Tasmanian Trades and Labour Council and the Health and Community Services Union, where she was elected assistant state secretary.

Harvey's parents. Photo courtesy of Ros Harvey.

Globetrotting: Europe and Cambodia

After Harvey was recruited by Public Services International (PSI) in 1998, the family relocated to Geneva. As executive officer she oversaw sixteen regional offices and over eighty staff members.

From 2002 through 2010 Harvey worked in Cambodia and Geneva, Switzerland, with various international organizations.

It was her stint as director of Better Factories Cambodia from 2003 to 2006 in Phnom Penh, however, that perhaps draws the closest parallels to The Yield. In 2003 Harvey was hired as the chief technical advisor of the United Nation's International Labour Organization (ILO) garment-sector working-conditions improvement project also known as Better Work. It was a critical juncture as the World Trade Organization's Multifibre Arrangement was due to expire at the end of 2004, putting factory jobs at risk.

Harvey and her team created a business model that used technology as part of the solution. International apparel companies were spending US$50 a worker annually monitoring factories. Harvey calculated that it could be done by the ILO at US$2 per worker at higher frequency and quality through data gathering and sharing.

She moved to Geneva with a "year's salary and a desk" to launch Better Work globally with the goal of taking it to all major apparel production countries.

"It was this idea of how do you create public good with private effort. . . . If you can improve working conditions, then firms that treat workers better were more resilient and more productive," she says.

A Start-up Is Born

Ros returned to Tasmania in 2010. Her young son Jules was seriously ill and the family wanted to be closer to home. On her return, Ros took up responsibility for the state's economic development strategy as deputy secretary of the Tasmanian Department of Economic Development, Tourism, and the Arts. At the time, Tasmania was to be the first Australian state that would have the Australian National Broadband Network (NBN) rolled out. It was the time when the government was looking to showcase the value of digital technology to business and the community.

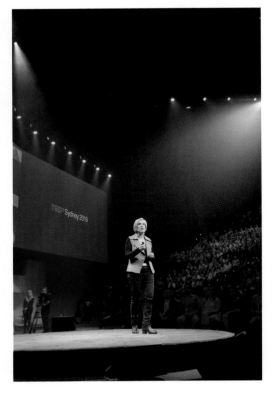

Harvey speaks at TEDx Sydney in 2019. Photo courtesy of Ros Harvey.

"I realized that I could easily transition the skills that I had learnt from one industry to another," she says, meaning from the apparel industry to agriculture.

In her new professional role she observed that roughly 30% of private sector production was in local agriculture. Further agriculture faced several challenges such as high labor costs, increasing input costs, the impact of climate change, and produce that had to be shipped offshore to get to market. She saw an opportunity to work with the research sector using digital technology to increase productivity. Harvey observed that agriculture had one of the lowest levels of digitization compared to other industries and saw a tremendous opportunity. She also points out that the world needs to produce 60% more food by 2050 to meet the demands of a growing population.

In 2012 Harvey took her broad idea of a social enterprise that would use innovation to help the agriculture sector to the University of Tasmania and named it "Sensing Tasmania," in short "Sensing T." She ran a small laboratory at the university and worked with development consultants, researchers, and farmers on a concept that in many ways mirrored Better Works.

Harvey's blueprint was as follows: the social enterprise would create a commercial model and collaborate with a research body (the university). Data such as water usage of crop yield that would

The Yield's sensor node (with rain bucket, anemometer, and barometer) located at Back Paddock Vineyard in Tasmania. Photo courtesy of The Yield.

help growers would be gathered, and at the same time the enterprise would build a data-sharing platform targeting researchers.

"We [thought] we would take the algorithms and commercialize them rapidly, because one of the biggest problems in research is that it is very slow. People take three years and then they get to the end and ask what are we going to do with this?"

The project fizzled, and Harvey called the partnership with the university a "mismatch."

"Universities are just not agile and responsive enough for what you need in an early-stage business," she says. Not one to give up easily, she made a rapid decision to take the concept and launch it as a start-up. By a stroke of luck she was also connected with Michael Briers, who is well known in Australia's IoT space. Brier had created something similar in the fintech space to what Harvey was planning in the ag space.

Frustrated with the amount of time it took data from academia to be shared with the market, he found a way to sell data to companies including Reuters and share it with researchers too.

"I was on a mission to create this ecosystem and have end users drive the research and not the other way around," Harvey says. Briers became a founding director of The Yield and was important in helping the start-up from its early days to its official launch in 2017.

Sensing+ Aquaculture, one of The Yield's first products for the aquaculture arena, uses sensors to measure a range of climate conditions from water salinity to wind speed to produce real-time data. The product helps oyster farmers avoid water contaminants such as the Pacific oyster mortality syndrome (POMS) virus that can wipe out entire harvests.

The Yield's team ultimately designed a system that offers a toolbox of technologies (depending on the customers' needs). Machine learning, for example, can forecast weather, yield, harvest volume, and someday quality too.

The Yield's key product is a subscription-based service Sensing+ that helps farms gather insight into the details of the microclimates on their land and in the process identify growing conditions on the farm. The information helps growers make decisions on prime times to irrigate, plant, spray, and harvest. The company's tagline is "Taking the guesswork out of growing." It affirms that the innovation can measure up to 2,500 different points over a 75,000-acre radius. Since 2017 some 350 customers have used Sensing +. Its target customers are multinational agriculture companies or big growers.

The team at The Yield connects over company activities such as painting.
Photo courtesy of The Yield.

From the very start, Harvey has made social responsibility a part of The Yield's mission statement. It encourages customers to make their data available to scientists and researchers who study the environment, ecology, or agriculture. In the company's description is a "mission to feed the world without wrecking the planet."

In recent years, Harvey has gained the limelight in the food, farming, and agtech space, whether through speaking engagements or media write-ups. She and Briers continue to collaborate as cofounders of the Food Agility Cooperative Research Centre (CRC) and the Knowledge Economy Institute, where she remains a strategic advisor.

The Yield was featured in Microsoft's high-profile global AI campaign in 2018

Harvey and her mother, Dr. Patricia Harvey, an emergency doctor. She says her mother is her inspiration. Photo courtesy of The Yield.

as a leader in how to use AI to solve global challenges. In that same year, Harvey also made the *Australian Financial Review*'s one hundred most influential women and was awarded the emerging leader in technology prize at the Women's Agenda Leadership Awards.

"I've always been good at raising money . . . it's as hard to raise a lot of money as a little bit of money, and basically it's not that different," she says. "It's about how you work with people—you've got to have a clear and compelling vision, a plan on how you're going to do it. People have got to believe you can do it, so you have to have a history of execution, and you've got to understand the flows of money and what motivates it."

But Harvey hesitates when asked what's next.

"I've got a long way to go yet," says Harvey, who states the long-term vision for The Yield: a global business that focuses on large agriculture corporations and a philanthropic arm that serves as a data repository.

"We want to be essential to five hundred of the world's biggest corporate growers within ten years," Harvey says, noting there are tremendous problems to solve such as the need for farmers to produce some 60% more food by 2050.

She lets out a generous laugh when asked whether there will be more start-ups or is this it for now. As she told *Produce Plus* magazine, if she weren't running The Yield she'd be a campaigner for climate change action, which she considers one of the most pressing problems on the planet.

Besides the various hats that she wears, Harvey enjoys spending time with family including her children, Jules and Sabine. She is a voracious reader via her Kindle, now reading the biography of Alexander Hamilton because she's a big fan of the musical *Hamilton*, which she's seen twice. "I am an economist and once maybe four or five years ago my daughter and I drove from Los Angeles to San Francisco before *Hamilton* became really mega, and she made me listen to it twice," Harvey says, laughing about the memory.

Her philosophy for life of "always trying to understand before being understood" holds true. Indeed it has been the North Star in her professional life and so far it seems to have taken her to some amazing places.

Citations

Tas Bindi, "Bosch, KPMG Back Agtech Startup The Yield in AU$6.5m Round," ZDnet, April 6, 2017, https://zdnet.com/article/bosch-kpmg-back-agtech-startup-the-yield-in-6-5m-round.

Kate Jones, "Ross Harvey Is an IT Entrepreneur with an Answer for Future Food Crisis," *Sydney Morning Herald*, November 17, 2017, https://www.smh.com.au/business/small-business/ros-harvey-is-an-it-entrepreneur-with-an-answer-for-future-food-crisis-201711160-gzmj5b.html.

Michelle Pelletier Marshall, "15 Minutes with . . . Ros Harvey, Founder of The Yield," Global AgInvesting, February 8, 2018, https://www.globalaginvesting.com/15-minutes-ros-harvey-founder-yield.

PENELOPE NAGEL

Cofounder and Chief Operating Officer, Persistence Data Mining Inc., San Diego, California

A Ninth-Generation Farmer and Open Water Swimmer Finds Her Groove as an Agtech Entrepreneur

Penelope Nagel is the COO of agtech company Persistence Data Mining. Photo courtesy of Penelope Nagel.

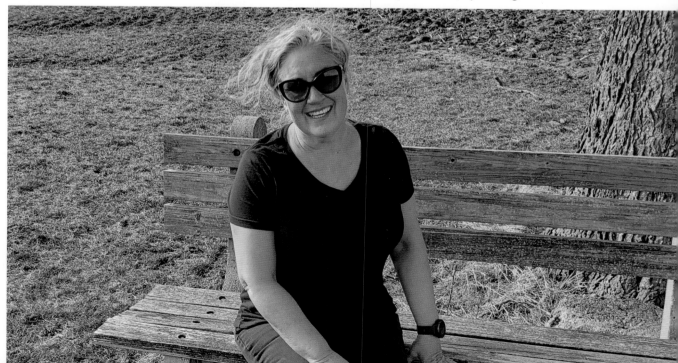

"Once you make a decision, the universe conspires to make it happen."

—Ralph Waldo Emerson

Leading an agtech start-up doesn't feel so scary when you have to swim next to manta rays and killer whales and have even encountered a great white shark. Penelope "Penny" Nagel is an avid open-water swimmer and chief operating officer of Persistence Data Mining Inc. (PDMI), a soil nutrient mapping and testing company that uses hyperspectral sensors to collect data sets with the goal of improving soil management. She is equally passionate when talking about swimming and start-ups. Numerous times Nagel has completed the Catalina Channel Swim, which is a

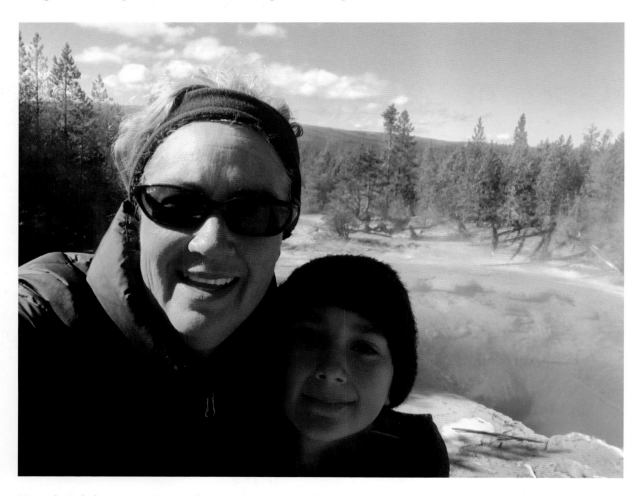

Nagel with her son, Peter, during a trip to Yellowstone National Park after conducting testing in Montana. Photo courtesy of Penelope Nagel.

Brian Zamudio, CEO of Persistence Data Mining, and Penelope at the R&D 100 Awards Banquet in 2019. Photo courtesy of Penelope Nagel.

little over twenty-two miles. This means swimming nonstop, often fifteen or more hours, through the night. Each swim brings a new challenge, whether it's the weather, currents, or sea creatures.

"A lot of these channel swims we do at night, and it's learning to face your fears, and get in there and do what needs to be done," says Nagel, who describes herself as a "channel swimmer." "There are moments during a swim when it comes down to one stroke at a time—it's just moving forward and doing the best you can with that next thing in front of you."

Marathon swimming is an example of Nagel's commitment to her passions. She brings the same level of commitment to her other passion, being an entrepreneur in a fledgling sector.

Her involvement in the launch and development of PDMI has also been a journey. In 2012 Nagel and entrepreneur Brian Zamudio launched the company with the goal of offering growers more detailed and less expensive soil nutriment mapping and testing data through PDMI's Soilytics subscription service. PDMI's target customers are the fertilizer companies that, at their own expense, often conduct soil tests for farmers, hoping to sell them fertilizer.

PDMI targets small- and midsize farmers, understanding and empathizing (Nagel herself is a farmer) with their challenges, including limited budget and resources.

"The number one issues for farmers is soil sampling—it's just too expensive, but if we increase the granularity, there are then huge differences," says Nagel, noting that the more samples taken on a farm or piece of land, the more information is elicited since land, depending on the size, can vary

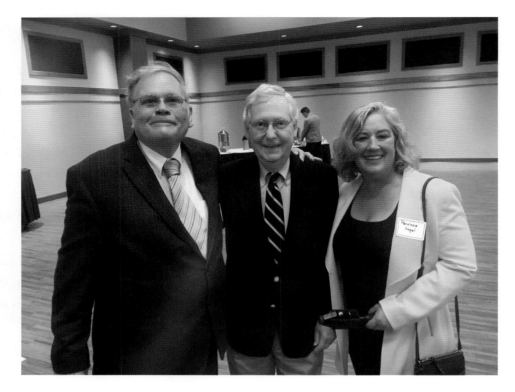

From left to right, Richard Proffitt, one of PDMI's investors, Senator
Mitch McConnell, and Nagel at the 2019 Lincoln Reagan
dinner in Kentucky. Photo courtesy of Penelope Nagel.

greatly in typography and even climate. She compares soil to skin. "Sampling is like a biopsy," she says.

The PDMI team built its start-up somewhat differently than other start-ups. Whereas fundraising is often at the forefront of any new enterprise, PDMI has pursued a strategy of first seeking acceptance from the agtech industry, academia, and agriculture.

Since launching, the start-up has been in active beta (testing) mode and busy working with individual farms and companies on paid pilot field trials. In such cases, the farmer pays for the testing and PDMI reports the results and validates the process with lab tests. PDMI has conducted a considerable amount of beta testing at Nagel's family farm in Mount Auburn, Illinois; since 2018 it has worked with agronomist Amy Gardner (who manages thirty thousand acres in Montana) on a field trial; and in 2020 it began to commercialize its product by partnering with a phosphorus company to pilot-test technology across seven states.

Nagel attributes the company's success in part to the accelerators it has participated in: AgLaunch in Memphis connected PDMI with a huge farmer network in the South for testing, and the THRIVE

Accelerator out of Silicon Valley helped connect the company with large agriculture companies and a farming network in the Salinas Valley.

Nagel also hopes PDMI will get buy-in from academia through collaborating with universities on projects or publishing in academic journals. In 2020 it published an article with the International Society of Optics and Photonics on new testing methodologies on organic matter detection and has also published with the International Society of Precision Agriculture. "It is important to get buy-in from academia because they are one of the go-to resources for agriculture," says Nagel. "We are also working with the USDA and United Nations Development Programme (UNDP) because they help make policies that affect decision-making tools in agriculture."

PDMI has collaborated with the agriculture industry on research too, such as with SoilHawk, which produces a robotic soil-sampling machine. Its work with the Ohio Soybean Council has paid off when the council and PDMI won the R&D 100 Awards in 2019. PDMI is also working with the UNDP to provide the company's soil analysis technology to farmers in developing countries.

As of June 2020 the company officially launched with its first commercial customer: Grangeville Agricultural Limestone Project, LLC (GALP), a regional lime company in Idaho.

Numbers Game

Nagel is the first to point out that it has taken PDMI perhaps longer than other start-ups to get to market, but she says the management purposefully took its time to fundraise and commercialize. She has made it a priority to get the industry's approval, including the USDA.

As of June 2020, PDMI raised just under $1 million through a variety of channels, including angel investors and grants. Nagel says the company has kept costs low by compensating most of its staff of seven primarily with equity, and also collaborating with agronomic service providers, whether laboratories or analytics companies. The partners come from their vast network of connections in the farming world.

Going into the start-up, "I knew it was going to be a hurdle and in agtech there is always a ramp up," she says. "I've had people ask me why aren't you asking for $5 million and I say we don't need $5 million. We kept costs real lean. We are trying to deploy something that is going to be profitable because it's real science and that it is the real thing. That might be a bad attitude to take, but we're not going to take $5 million so we can say we have a series A round."

When it comes to raising money, PDMI has turned to individual investors and grants. In the spring of 2020, Nagel was preparing a National Science Foundation grant to submit to various colleges, including West Hills College in California and Murray State University in Kentucky, as the colleges look to step up their agtech programs.

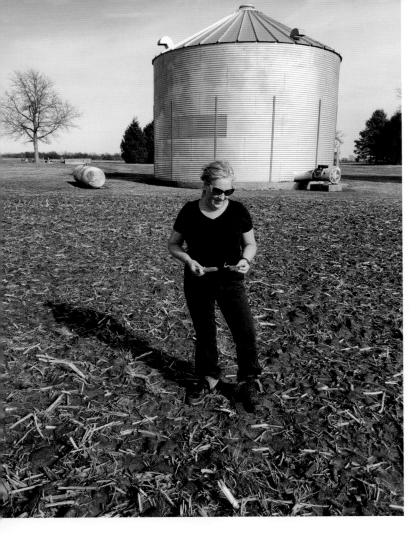

Nagel grew up on a family farm in Illinois. Photo courtesy of Penelope Nagel.

Nagel drives tractors, her children grew up on the family farm, and she spends a lot of time in the fields. She is able to comfortably navigate the two very different worlds that make up her business: one day she was having dinner with movers and shakers such as Senator Mitch McConnell (R-Kentucky) and the next day she was soil-sampling with growers.

"It's a whole different world when you're standing out there seeing how much ground there actually is. Unless you are standing out there sweating, it is just numbers," she says. "I do see a lot of people and they don't have their feet in the soil. It's been commented to me that people creating these technologies aren't really invested in being out in the fields. They have great connections and are able to raise a lot of money, but it's difficult watching people raise a lot of money when you realize the farmer isn't going to buy it in the end."

Nagel has gone out of her way to seek a stamp of approval from the agriculture industry, especially the National Resources Conservation Service (NRCS), an agency of the USDA, to show PDMI's testing method "is legitimate science and not just hocus-pocus."

PDMI's focus is now to attain the American Society for Testing and Materials standard, a required step in receiving ISO certification from the International Organization for Standardization (ISO) that shows a company meets quality with its management.

Farming Roots

Although she was born and raised in San Diego, California—where PDMI is located—she defines herself as a ninth-generation farmer. Farming is part of the family lineage: her maternal ancestors settled the ground in Illinois in 1818 and farmed for a living. She continues to oversee the family farm of somewhat fewer than a thousand acres of corn and soybeans.

"My memories are [of] returning to see my grandparents and great-grandparents," Nagel says. Neighbors, many of them fellow farmers, fondly recall her grandmother's cherry pie. Since starting PDMI, she's created a barter system with local farmers for soil testing that involves cherry pies and biscuits and gravy.

Nagel decided to follow her interest and ambitions in finance and majored in economics at San Diego State University and later earned her MBA at the University of Phoenix—but a series of events would lead her back to agriculture. First, in 1991, while in college, Nagel was in a motorcycle accident that left her in a wheelchair for five years; eight years of reconstructive surgery on one leg followed. This unfortunate twist of fate did lead to one happy result: her love of simming. "It's the only sport I can do," she says.

Nagel with agronomist Amy Gardner in the lower valley in Kalispell, Montana, during their first year of testing in Montana in 2018. Photo courtesy of Penelope Nagel.

Her maternal grandmother died a year after the accident and Nagel's mother inherited the farm. This also meant Nagel stepped up her visits to the farm to help her mother and two sisters. Every spring she would conduct an extensive review of the farm's budget, and it struck her that there was a close connection between the budget and the decisions farmers make about everything from fertilizer to sprays. The cost of fertilizer, for example, was soaring in 2009 due to increased demand.

Many small- and midsize producers struggle with balancing soil testing's cost with its benefits. Soil sampling can be very expensive, but the more samples obtained the more extensive the resulting analysis can be for farmers, leading, in theory, to their making better decisions on their crop plan. When Nagel came up with the idea of PDMI's agricultural service, she had the small farmer, as well as farmers in emerging countries, in mind.

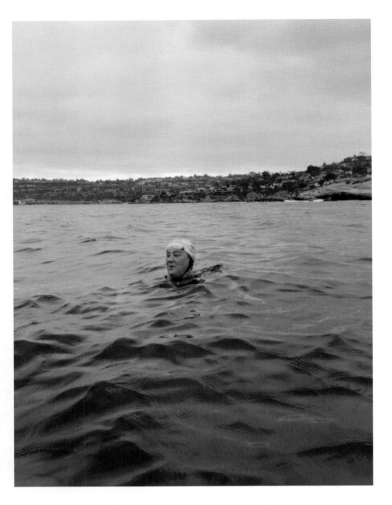

Nagel in July 2020 during an eight-mile training swim from La Jolla to Belmont Park in Mission Beach, San Diego. Photo courtesy of Penelope Nagel.

"I saw a need for better input decision-making tools," she says (meaning implementing new tools that would enhance soil analytics to make better decisions).

"Most solutions were reactive and the only proactive method was lab testing, which was expensive and therefore not done on a granular-enough level to help profitability or the environment."

The next turning point came in the decade she worked in finance with banks such as HSBC and Santander. At HSBC, in 2005 she was one of a select number of employees chosen to participate in an Earthwatch project. Nagel had the chance to go to Iceland and work with scientists to research water flows from the glaciers.

"I loved what I was doing and I felt like I was making a difference," says Nagel of the expedition.

The real aha moment came when she was laid off by the bank in a 2011 down-sizing. Brian Zamudio, a friend from her college days, invited Nagel to work on his start-up, Persistence Data Mining, with him and she accepted the offer. The company's testing methodology was being used for mineral exploration and for the mining industry. A few years later, as the oil and gas industry began to wane, Nagel suggested shifting sectors.

Their testing methodology/innovation, she pointed out, could be used to examine micronutrients in the soil, including nitrogen, phosphorous, and potassium, and thereby provide information to help growers make input decisions.

"I kind of look at the soil like it's a living thing and it has to have the right balance, just like the food that goes into our body. If it's [nutrients] not going into the soil, it's not going into the food and it's not going into us. It starts with the nutrients in the soil so we can all be healthy," she says.

Nagel at the family farm in Illinois. Photo courtesy of Penelope Nagel.

The Future

By late spring of 2020 PDMI, like all other businesses, had to wrestle with COVID-19 and brace itself for the pandemic's impact on business and financials. The team's travel plans to set up new customers were stymied and customers involved in pilots have had to mail in soil samples.

However, in the eight years and counting since she's been in agtech, Nagel says she's found the journey fulfilling, even with all of its ups and downs. Home life also keeps her busy with a son entering eighth grade and a daughter in college. She's also tried to "pay it forward" by mentoring a young woman who dropped out of high school. She took the young woman to industry conferences and to Mexico, where they volunteered at an orphanage based on a farm. With Nagel's help the young woman applied to college and ended up getting into the American University in Paris. Nagel recently received a call from her protégée, who delighted her with the news that she was studying in the department of agriculture.

"She said you'd think it was the universe that stuck me here after all those years with you and I said go with it," Nagel says. "The opportunities are endless and if you know anything about agriculture you will be in a better position to feel rewarded by that job. I always say the most noble position in the world is that of a farmer: [a] farmer feeds people, protects the environment—there's family values associated with it."

Swimming in open water and running a start-up, especially in a fledgling sector, draw similar parallels, Nagel observes. She finds the ocean a safe haven and a place that energizes her. Her typical day, no matter how busy, usually includes a quick one- to two-mile swim in the ocean. In a typical year she logs about 650 swimming miles.

The marathon swims are also a way for Nagel to give back to a cause near and dear to her heart, raising money for Project Mexico & St. Innocent Orphanage. She raises money from sponsors and supporters for these swims and donates the money to the orphanage.

She compares swimming to working as an entrepreneur.

"Working on a start-up is a marathon and not a sprint; it's a journey, it's a quest, and it's a process. It's being able to stay the course when things aren't going your way," she says. "We have successes and failures and we keep moving forward."

Swimming long distances in open water has also taught her to have faith. One day, Nagel and her training partner were on a practice swim when they saw a great white shark. At the same time a pod of dolphins swam to the women.

"There's a fate aspect to this, and they circled us for an hour and got us back," she says.

And when rough waters come her way, she tries to remind herself of her passions and inspirations—"love of food, the ocean, and the land. Being part of this world and a part of the solution to big problems inspires [me to] move forward toward a better future."

OTHER TRAILBLAZERS IN AGTECH

Agtech also crosses over with the burgeoning subsectors of agbio and foodtech, so it was no surprise that during our search for women founders and leaders in agtech we came across dozens upon dozens of inspiring stories of women leaders in the larger arena of the food production industry. These women include a fieldworker turned farmer, a farmer turned banker, and a policy maker and consultant who transformed into an agtech pioneer in her home state.

We are sharing their stories too in the spirit of communicating that agtech is not a sector in a bubble. It includes a vast range of professions and opportunities in food and farming. Everything is connected and interrelated when it comes to the production of food.

ALLISON KOPF

Founder and CEO of Artemis, Brooklyn, New York

Allison Kopf reached a crossroads in 2011. She had been working in the solar industry, but the industry was suffering. Even when an investment bank offered her the opportunity to invest in commercial solar deployments, she realized it wasn't exciting for her. She had studied physics at Santa Clara University in California and had built a solar-powered home for an international design-build competition, but she was far more interested in technology development than working in finance.

Instead, Kopf joined BrightFarms, an early-stage start-up with a business model that combined the development approach of solar with controlled environment agriculture.

By 2015 Kopf had spent four years working as the real-estate and government relations manager for the company, one of the fastest- growing indoor growers in the US. This experience whetted her appetite to explore other options for working in agriculture. "Originally I fell into it and then I fell in love with it," Kopf says of agriculture. This was the first time that Kopf, who grew up in the suburbs outside New York City, had worked on a farm.

Farming did not run in the family, but entrepreneurship did. Kopf grew up in Rockland County, New York, where her father worked as the commissioner for finance for the county for thirty years while running his own accounting practice. Her mother is the treasurer for their village and also the founder of a travel blog named *The Open Suitcase*, which attracts millions of readers.

Upon joining BrightFarms, Kopf says, "I became obsessed with farmers and how they ran their operations." Moreover, she saw opportunities in the industry, especially when it came to the supply chain and processes from production to distribution. "Agriculture is one of the few industries that isn't going anywhere, it will always be around, and it has one of the most innovative and yet least innovative supply chains in the world." The solution to problems, including lack of data, was creativity.

Although many industries were seeing rapid advancements in technology—the self-driving car, for example—"The digital layer was missing in agriculture," says Kopf. "It's incredibly similar to manufacturing and other industries, where the introduction of software has led to massive industry transformation. Without software, the industry lags, and it's incredibly frustrating for the farmer. I was personally frustrated with the lack of agriculture-specific technology available."

That frustration resulted in her decision to leave BrightFarms and build Artemis (originally named Agrilyst). She teamed up with Jason Camp to launch the company and create its flagship product, the Cultivation Management Platform (CMP). The CMP is an operational software platform for growers to help them gather data and manage their operations. In the beginning the company primarily targeted indoor farmers and greenhouse operators. The endeavor was in many ways a combination of her two passions—technology and agriculture.

Out of the starting gates, the company sought customers by tapping into their own networks and hitting trade shows. Artemis experienced early success on the funding front, including backing from Brooklyn Bridge Ventures, which led the company's seed round of funding. In 2015 Artemis was one of twenty-five startups chosen from a pool of thousands to compete in TechCrunch Disrupt SF, an annual conference that brings together investors and entrepreneurs in technology. They won the event's competition. The company also participated in the Pearse Lyons Accelerator, run by Alltech.

Allison Kopf is the founder and CEO of Artemis. Photo courtesy of Allison Kopf.

Artemis is subscription-based software that helps specialty crop growers run more efficiently by letting them manage everything from seed through sale, including crop schedules, task management, food and farm safety, labor needs, reporting, metrics tracking, and inventory management. In recent years the customer scope has included cannabis growers too. Although the company's key markets right now are Canada and the US, the software is available in more than ten countries.

Kopf (right) takes the stage at the Forbes AgTech Summit in Salinas in 2019.
Photo provided by Allison Kopf.

Kopf also supports other female founders through XFactor Ventures, which invests in early-stage companies led by women. She feels strongly about paying it forward, and in a 2019 article she wrote for Medium's *Startup Magazine,* she asserts that although diversity is critical in hiring, it is equally as important for boards.

"This isn't just about creating an inclusive team, it's about choosing partners who value diversity as well. The Artemis board is led by women—75 percent of our directors are women," she wrote.

In an interview in March 2020, Kopf said, "I am focused on helping the next crop—excuse the agriculture pun—of female entrepreneurs build billion-dollar businesses."

As of May 2020, Artemis had twenty-seven employees and had raised a total of $11.7 million. Headquartered in Brooklyn, New York, the team has locations across the US and in Canada and Chile.

When not working Kopf spends her time reading and learning about how others built their businesses, as well as working out. "I've run two marathons and it's probably time for another soon," she says matter-of-factly.

ANDREA CHOW

Senior Vice President of Engineering at Ontera, Santa Cruz, California

Andrea Chow was born in Guangzhou, a city in southern China, and grew up in Hong Kong. When Chow was a teenager her family immigrated to the US and settled in Los Angeles, California, where her father continued to run the Hong Kong bakery and her mother worked as a seamstress. Chow is a chemical engineer by training at the University of Southern California. She later received a PhD at Stanford University. She calls her journey into the biotech and agtech space a "little windy" because of her chemical engineering background.

"I remember when I was in high school, taking a biology class, and the teacher was not at all inspirational, and I told myself I would never work in anything related to biological science. Biology seemed to be a lot of rote memorization, and that was not very easy," she says, laughing. "Then I

Andrea Chow of Ontera explains that the sample-to-answer platform called NanoDetector uses a small cartridge (about the size of a smartphone) to automate the laboratory workflow of sample amplification and nanopore measurement. Photo courtesy of Benji Hsu.

did my undergrad and grad schoolwork in chemical engineering. I focused on a field called polymer rheology (polymer material flow and processing), where the applications are broad."

She started her professional career in the defense industry and worked as a staff scientist at Lockheed Martin's Advanced Technology Center in Palo Alto for eight years, but she says she "aspired to a higher goal." After the NASA project that she was working on—developing a new solid rocket motor—was cut, projects related to building missiles for defense did not fulfill her passion. She wanted to devote her time to endeavors that she viewed could directly make the world a better place.

She fell into biotech when she joined Caliper Technologies Corp., a company based in Palo Alto, California, that focuses on an area called microfluidics, that is, driving fluid at a very small scale in order to automate biochemical analysis for research, drug discovery, and clinical testing. The company was looking for engineers with biotech experience. "They [Caliper] took a chance on me, so that's how I got into biotech," she explains. She beefed up her biology background by taking extension courses in molecular biology and cell biology at UC Berkeley. Along the way she found that "the reason I had actively sought to go into biotech was the exciting advances in genetic engineering that were going on at that time. I knew that biotechnology was a very rich area to explore."

Chow was also entrepreneurial, founding and running her own company, Caerus Molecular Diagnostics, from 2009 to 2013. The company's goal was to develop low-cost and high-accuracy DNA sequencing technologies. She raised half a million dollars from government grants but later left the start-up, as it was very challenging to raise additional capital during the Great Recession. Next she led the engineering efforts at Promega, first, and then BioElectron Technology Corp. in Silicon Valley.

In February 2019 Chow joined Ontera, a biotechnology company that had just begun expanding into the agtech space. The 13,000-square-foot space in Santa Cruz houses some thirty-five employees. While the company doesn't disclose details on capital raised, its series A was led by Khosla Ventures. Ontera's products are based on novel solid-state nanopores, which are sensitive biosensors with nanometer-size pores fabricated in semiconductor materials. The nanopores are capable of counting and classifying single molecules of nucleic acids and proteins. As part of its efforts to expand into the agtech space, in early 2020 Ontera launched the NanoCounter for research applications and is currently developing a sample-to-answer platform called NanoDetector, a point-of-care molecular diagnostic system that can rapidly quantify genetic traits of seeds and plants and detect plant pathogens.

ELIZABETH "BETH" BECHDOL

Formerly president and CEO of AgriNovus in Indianapolis, Indiana

Beth Bechdol hails from the cornbelt of the US. She was born and raised on the family farm in Auburn, a largely rural community of some fifteen thousand in northeastern Indiana.

Agriculture is very much a part of her personal history. She comes from a seven-generation family of farmers in Indiana, where the family farm specialized in row crops, including corn, soy beans, and wheat. As expected in many traditional family farms, her father began working on the farm as a boy and later took over managing it.

Beth addresses national FFA students on the FFA blue stage about how they can lead change to feed the world, protect the planet, and improve lives. Photo courtesy of Beth Bechdol.

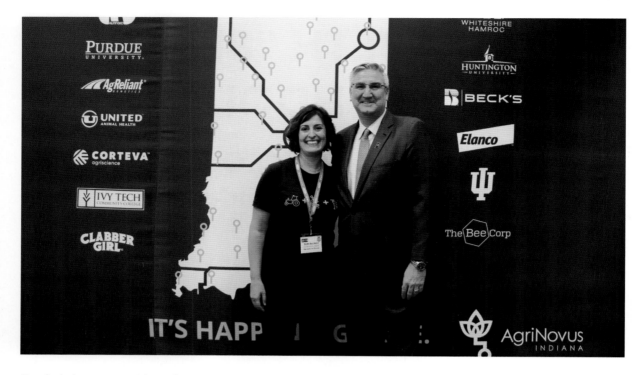

Bechdol meets with Indiana governor Eric Holcomb. Photo courtesy of Beth Bechdol.

"You know in farm families, at one point it was kind of like a monarchy. You needed a son to take over the family farm, and he grew up knowing that he was going to take over the farm because this was the plan for him," says Bechdol, forty-six.

She continues: "I give my dad a lot of credit. He was a Future Farmers of America [FFA] officer in Indiana and super smart and super accomplished in 4H and the FFA." Bechdol and her sister (eleven years her junior), however, had other interests and aspirations. Bechdol had dreams of leaving Auburn and pursuing a career in international affairs. Looking back, she says she feels blessed that her parents didn't pressure her to stay on the farm.

"I didn't want to be a part of it. I wanted to go to a big city and I wanted to be a part of all of the music and academic programs in school. I was part of the international speech team, I was in the band, I played tennis," Bechdol recalls. "My dad was like, 'You need to go down the path you need to go down.' So, for me I started down this path of being interested in the world."

At Georgetown University she found that she could have a career in international affairs *and* agriculture, the latter of which was still near and dear to her heart. After graduating, Bechdol spent a decade in Washington, DC, working in policy consulting and roles on the US Senate Committee on Agriculture, Nutrition, and Forestry. She was also a member of the special projects staff for the American Soybean Association.

"The first chapter of my career was all policy: it was farm bills, trade negotiations, and market access with foreign countries," Bechdol says. But she, her husband, and their young daughter returned to Indiana after her mother became terminally ill. Although she had intended to stop working and be a stay-at-home mom, little did she know there were other plans in store for her.

It was in Indiana that she found her true calling in working as deputy director of the State Department of Agriculture under Governor Mitch Daniels and in leading the agribusiness sector at Ice Miller LLP. Most recently she was involved in the launch of AgriNovus, which would put her on the agtech map. AgriNovus, started in 2017, is dedicated to developing agbioscience in Indiana.

At the end of 2019, Bechdol left AgriNovus to take on a new role as deputy director-general of operations at the Food and Agriculture Organization (FAO) of the United Nations, in Rome, Italy. The FAO is a UN agency focused on world hunger.

CLAUDIA PIZARRO-VILLALOBOS

Director of Marketing and Culinary at D'Arrigo Brothers, Salinas, California

Claudia Pizarro-Villalobos was born and raised in Salinas, California. She's the head of marketing at D'Arrigo Brothers, one of the largest producers of broccoli rabe (or rapini) in the country. The family-owned grower, packer, and shipper, based in Salinas, grows and sells broccoli rabe, fennel, broccoli, cauliflower, and romaine hearts under its flagship brand Andy Boy, named after Andy D'Arrigo. The produce can be found in supermarkets and stores across the U.S. The company has 1,700 employees combined in Salinas and Yuma, Arizona.

Pizarro-Villalobos grew up the daughter of parents who emigrated from Mexico to the US. Her mother worked at a hair salon and her father at the Spreckels sugar factory before they opened a Mexican restaurant named Chapala in Salinas. After high school, Pizarro-Villalobos attended the

Claudia Pizarro-Villalobos is the director of marketing and culinary at D'Arrigo Brothers.
Photo courtesy of Claudia Pizarro-Villalobos.

University of California, Berkeley, where she majored in ethnic studies. She later earned her master's degree in higher education at Harvard University.

I first met Claudia when I was reporting on agriculture for the *Californian*, for which I wrote a number of stories about or related to D'Arrigo Brothers. Claudia and I also led a special panel together at EcoFarm 2020 on how to use social media to effectively tell stories and boost one's brand.

Here she talks about the journey that led to her job and what inspires her to continue with the work.

Q: What led you to the agriculture sector? Does ag run in the family?

A: When my parents emigrated from Durango, Mexico, they did not speak English, have a formal education, nor much money to get by. They landed in Central California as they were told about farm-working opportunities there, and both worked in ag for many years. My mother wanted

A typical day for Pizarro-Villalobos involves working in the fields and in the office at D'Arrigo Brothers' headquarters in Salinas, California. Photo courtesy of Claudia Pizarro-Villalobos.

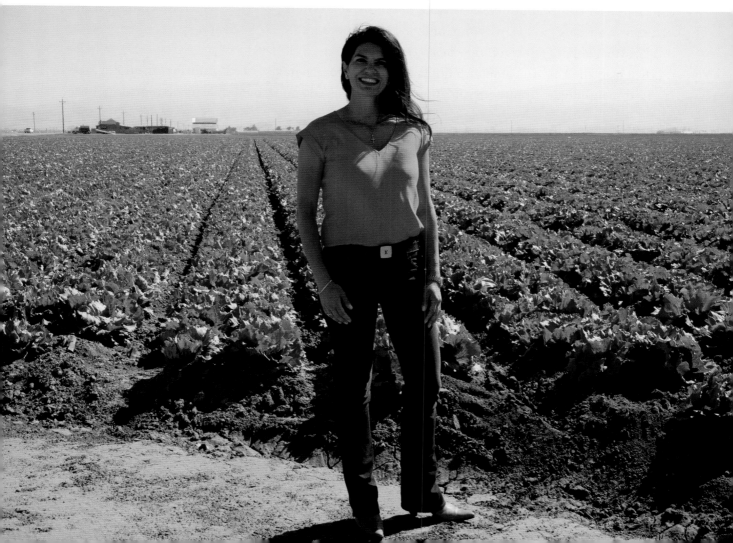

a change due to the laborious nature of farmwork and took classes at the Salinas Adult School to learn English and began to cut hair at a salon. My father began to work for Spreckels Sugar Company and got by with very little English. My mother and father opened a Mexican restaurant named Chapala in Salinas, and it thrived as a family business for twenty-five years. The business was the livelihood of our family and put my siblings, Sandra and Ruben, and myself through Catholic schools (kindergarten through high school) and college (undergraduate, law school, and graduate school). So, you can say that my family's humble beginnings started in agriculture and over the years we worked hard to attain the American Dream.

Q: Through your work and what you do, what are your main contributions to your company?

A: I am proud of the broccoli rabe marketing campaign, creation of Andy Boy recipes, culinary videos, and launching the Sip, Bite and Turn Up the Pink event, which raises awareness and funds for the Breast Cancer Research Foundation. The broccoli rabe campaign enlisted a celebrity chef, nutritionist, and an array of culinary influencers to heighten the versatility, cooking techniques, recipes, and nutritional attributes of broccoli rabe. On June 15, 2016, the *Wall Street Journal* ran a feature entitled "Broccoli Rabe Dreams Big," and it underscored the success of the broccoli rabe campaign. We continue to build our library of recipes and culinary videos to educate customers and consumers on how to cook with our Andy Boy commodities.

Q: What inspires you to continue what you do?

A: I have the honor of being a bilingual storyteller for a family-owned company that is celebrating its one-hundredth anniversary in 2020 and has a trusted and recognizable pink-colored produce brand (Andy Boy) that is revered nationally and internationally. I thoroughly enjoy the fact that in marketing no one day is the same and I can communicate in Spanish and English to reach a broader audience. One day I may be working on social media, culinary videos, recipe photos, setting up promotional collateral for retail ads, updating the website, working with influencers, filming farmworkers harvesting in the field, and the next I may be creating produce spec sheets, preparing slides for a board of directors meeting, coordinating a company holiday luncheon, customer appreciation event, or serving as an ambassador for the company.

Q: A quote that epitomizes you or inspires you?

A: "Change will not come if we wait for some other person or some other time. We are the ones we've been waiting for. We are the change we seek." —Barack Obama

DEB CASURELLA

Founder and Chief Executive Officer at MyAgData, Effingham, Illinois

Deb Casurella considers the Midwest her stomping grounds. A native of Chicago, she grew up in the city's South Side and now lives in Wisconsin. She earned her bachelor's degree in business administration and master's degree in training and organizational development from the University of Minnesota. Her only living grandfather built locomotives for General Motors and one grandmother worked as a keypunch operator for *Time* magazine and was an Olympic qualifier in swimming many decades ago.

"There is no agriculture in the family, but I very much enjoy real people, real problems. Farmers are real," Casurella says. "They care for each other, their communities, and work to protect their families. They respect their history and ensure the family farm is there for generations."

After college, Casurella worked in the health care industry with stints as chief information officer at Definity Health and Prime Therapeutics. Her forte was leading or implementing information technology, and for some twenty years she also ran Casurella Consulting, where she helped organizations implement technologies to improve their businesses.

Deb Casurella is the CEO and cofounder of MyAgData, which takes data collected from sensors during planting and uses it to streamline reporting of acres to the USDA. Photo courtesy of Deb Casurella.

Casurella mows the lawn to help out a farmer while he was out in the field planting corn. "MyAgData team members were in the cab so it was a way for us to help him out while he helped educate our team members on planting and data collection during the planting operation," Casurella says. Photo courtesy of Deb Casurella.

Casurella became involved in agtech in 2009 when she was named senior vice president and chief information officer of ProAg, a global crop insurance company. In 2012 she founded MyAgData with the goal of simplifying producer reporting for crop insurance and government farm programs. The software system takes data collected from sensors during planting and uses that to streamline reporting of acres to the USDA. Target customers include producers, crop insurance companies, crop insurance agents, ag lenders, and ag service providers.

"It's similar to the IRS connecting to my bank to get the amount of interest paid in a tax year. It makes it easier to complete required reporting without having a stack of paper and handwritten notes," she explains.

While COVID-19 has created a number of challenges for MyAgData, including a disruption of food systems, Casurella says the pandemic has been somewhat of a silver lining for her company. On the company's website is the tagline "Keep your social distance and report your acreage from home with MyAgData. We don't change what you report, we change how you report it—electronically."

"The idea of electronic reporting is gaining traction as the months progress," she says. What will be needed to maintain a positive momentum for the ag tech sector are patient investors and policy changes in Washington. "The USDA has piloted electronic reporting and intended to go live this fall. We need this administration to proceed on that track," she says.

As of June 2020 MyAgData, with sixteen employees, has raised some $4.5 million. Despite the challenges that come with running a start-up she says what inspires her are "real people, real problems, and farmers are as genuine and real as you can get."

Outside of work Casurella's passions include hiking, golf, and underwater photography. She is also an active sailor.

Q: Have you faced any challenges as a female leader in the ag industry and if so, how did you overcome them?

A: My roots are in information technology, which like agriculture, is a more male-driven industry. But there is always room for women to shape the face of the ag industry. I'm confident in my ability to lead and [in] my industry knowledge. If I don't know something, I am able to use the people and resources around me to mitigate any challenge I may face.

Q: Why do you believe it's important to simplify the acreage reporting process?

A: Many farmers have newer and more robust technology than [do] crop insurance agents or staff in government offices, and I believe it's time to evolve from a manual data entry–driven environment to [a] more digital, customer service–oriented world.

Note: A version of this interview first appeared in "MyAgData Leader: Getting to Know Deb Casurella," September 9, 2019. See myagdata.com/2019/09/01/myagdata-leadership-getting-to-know-ceo-and-co-founder-deb-casurella/.

ERICA RIEL-CARDEN

Global Capital Markets Inc., San Francisco, California

Erica Riel-Carden is a passionate grower who is also an agtech attorney and investment banker. She's a principal at Global Capital Markets, a California-based boutique investment bank, working out of its San Francisco office. At Global Capital Markets she structures transactions for companies seeking to fundraise from institutional investors. She's helped sensor, software, biotech, and blockchain technologies enter the food supply chain, and the process has raised over $25 million for companies in the ag and agtech spaces.

Riel-Carden is also active on the agtech speaking circuit: from 2016 to 2020 she spoke at over forty events, which (until COVID-19) made her a frequent traveler. The conferences, summits, and roundtables have served her as platforms to connect with potential clients and also to scout out promising start-ups. Before joining Global Capital Markets, she had launched her own consulting company that advises agtech start-ups. Prior to that, she was an incubator attorney with Royse Law Firm's Silicon Valley AgTech Group. Riel-Carden's website perhaps

Erica Riel-Carden speaks at the SDL (Sustainability & Digitalization Leaders) Executive Leadership Summit Commercializing Change in Agtech in Miami in 2019. Photo courtesy of Erica Riel-Carden.

Riel-Carden served as a coach at the International Food and Agribusiness Management Association (IFAMA) Symposium, summer 2019, in Hangzhou, China. Photo courtesy of Erica Riel-Carden.

best and most concisely defines what she does: "I help companies and I know how to make plants grow too."

For Riel-Carden, agriculture is a personal passion; it also runs in the family history. Born at the Maxwell Air Force Base in Montgomery, Alabama, she grew up in Fairview Heights, Illinois, a suburb in southwest Illinois. A first-generation Filipino American, she has relatives in the Philippines who currently own and run farms. Her great-great-grandfather Francisco Baisas was a renowned entomologist and is credited with helping to control malaria worldwide. Baisas's research and writings are archived at the Smithsonian Institution in Washington, DC, and at the Field Museum of Natural History in Chicago.

When Riel-Carden was a child her mother started working for a large US agricultural company, where she leads research-and-development initiatives, commercial pipeline reporting, and global governance processes. Her father spent seven years in the US Air Force and continues to work in computer network operations and engineering as a government civilian worker. Her parents divorced when she was a child; thus Riel is her father's name and Carden is her stepfather's name. Her stepfather works in manufacturing, transportation, and logistics.

"Agriculture and technology was not something we focused on during any family meal. I just lived in it and around it. My hometown has an average of fifteen thousand residents and my neighborhood is still surrounded by fields of corn and soybeans," says Riel-Carden matter-of-factly.

At Ohio State she started pre-law and was an English major with the goal of attending law school, but after attending Horticulture 101 she immediately switched majors to agriculture. With the idea of working in agriculture, she also minored in Spanish.

"It just made me happy. One of my favorite childhood memories is remembering being outside while either my mother or grandmother worked in the garden," she says. Her first job out of college was working on plant germplasm research for the Ornamental Plant Germplasm Center, a joint effort of the US Department of Agriculture, Agricultural Research Service, and The Ohio State University. Soon after she was hired as a grower at a premium Midwestern wholesale nursery, where she also oversaw plant production on seventy-five acres of ornamental shrubs; among her responsibilities was supervising and translating for Mexican guest workers.

"Then life quickly pulled me out to California," she says. There Riel-Carden returned to school and earned her law degree at Santa Clara University School of Law. Her long-term plan was always to become a corporate attorney for an agricultural company and eventually to her own small farm and greenhouse.

Outside of work, running, yoga, and gardening keep her busy and her two toddlers keep her on her toes. "They're still young enough to pick up and take off on a moment's notice," she laughs.

JACKIE VAZQUEZ

Director of Operations at Sierra Farms, Moss Landing, California

Jackie Vazquez spends most of her workday in the fields, walking the vast landscape of strawberries that are bookended by the Pacific Ocean and the cypress trees that the Central Coast is known for. She's examining the quality and quantity of the berries and also connecting with the crew, including fieldworkers, as she walks.

As head of operations for Sierra Farms, A Good Farms Operation (Good Farms is a strawberry grower in Moss Landing), she oversees roughly a dozen farm managers who, depending on the season, manage a crew of more than 1,200 production staff. She's charged with some four hundred acres, which, depending on the time of year, grow a mix of strawberries and raspberries. The work is year-round and intense, as the farm (both a grower and a shipper) is known for supplying major retailers such as Costco and Kroger.

Although there's no official work attire, in many ways, she fits in as one of the guys on the farm with her jeans, vest, plaid shirt, boots, and hat to shield her from the heat and sun. Vazquez is an

As the director of operations, Jackie Vazquez spends a significant part of her day in the fields managing the crews. Photo by Amy Wu.

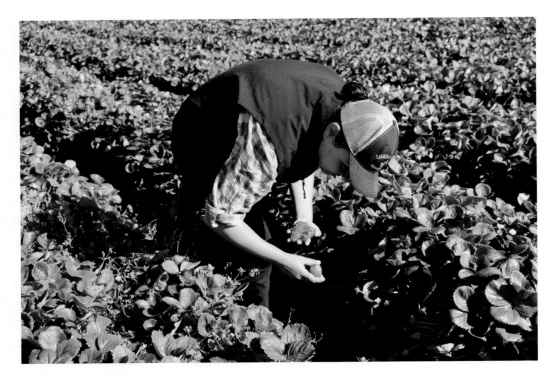

Strawberries are the flagship crop at Sierra Farms. Photo by Amy Wu.

anomaly, one of the few women who head operations at a midsize farm. It is a dirt and boots sort of job, where she juggles both business and farm operations.

Vazquez was born in Patterson in the Central Valley, California, and raised in Watsonville, a city north of Salinas known for its strawberry industry and annual Strawberry Festival. Her parents went from being farmworkers to running a chain of grocery stores (the one remaining store is La Esperanza Mercado), where most of their customers were farmworkers. Vazquez dropped out of the University of Chicago, where she was studying business and marketing, after a family emergency. She fell into the agriculture industry after accepting an administrative assistant job at a farm, which is where she came to learn more about the ins and outs of farming. As of 2020 she'd been in the agriculture industry for fourteen years, the six most recent at Sierra Farms.

When not on the farm, she spends a considerable amount of time with family, including her four sisters and father (her mother passed away in 2016 of breast cancer). She and her husband, a graphics designer, are hands on in raising their two children, ages ten and seven. Vazquez is a member of the PTA, and during the COVID-19 pandemic she and her husband split shifts in homeschooling the children.

She jokes and says she and her husband are polar opposites. "He is very much the creative one and I am very much the one who is two plus two equals four," she laughs. "I am black and white with gray and my husband is every color."

Q: What is your main job?

A: Being out in the fields and looking at the quality of our produce with our managers, and seeing that our crews are out doing what they are supposed to be doing. I am doing estimates for sales, and then managing the profit/loss margins and looking at the budget part of operations. I am making irrigation decisions with our input manager, meaning, hey, do we need to invest in more technology, what kind of irrigation meeting with scientists is needed if we have issues on the fields—so basically all aspects of growing and all aspects of running the business would be under my responsibility.

Q: What's your outlook for the future when it comes to women in agtech?

A: Right now, there's more of a push to talk about it [the gender disparity in agtech], but I think it's because there's more of a push for women in everything—it's just agriculture's turn right now. I definitely see more and more women graduating from Cal Poly and UC Davis and a lot of women I see going into ag are still going the food safety route and agscience, so I definitely see an increase of women in ag, but I haven't seen a big growth of on-the-farm, boots-on-the-ground women in agriculture.

Q: What inspires you to do what you do daily?

A: The idea of what we do is really feeding the world—that you have a direct impact on the consumer every single day you meet a basic need, which is very hard to do. Right now, for example, we are in a different time [referring to COVID-19], but right away farmworkers were considered essential regardless of the politics. Doing that work daily and knowing that product goes to a family—that just does something to me.

JOANNE ZHANG

CoFounder and CEO, Phytoption LLC, West Lafayette, Illinois

Chinese medicine seems like a distant link to farming, but for Joanne Zhang, its connection to health and wellness led her to foodtech. Zhang's grandfather was a doctor of traditional Chinese medicine (also known as TCM) who was well respected in the field.

"My grandfather saved many lives during World War II and treated many patients during the hard times following. Unfortunately, he passed away when he was still relatively young—decades before I was born," says Zhang, saddened that a lot of his wisdom was not documented and thus not passed to future generations. That said, her grandfather made a significant impression on her, ultimately inspiring her to work in the foodtech arena, defined generally as the intersection between food and technology.

Growing up in China, Zhang was fascinated with Chinese medicine. She says that it was "not only associated with skills in diagnosis and treatment, but also utilizing natural herbs and remedies to cure illnesses and nurturing the people's wellness." She's also a strong believer of traditional Chinese medicine's mantra of "Food is the best medicine."

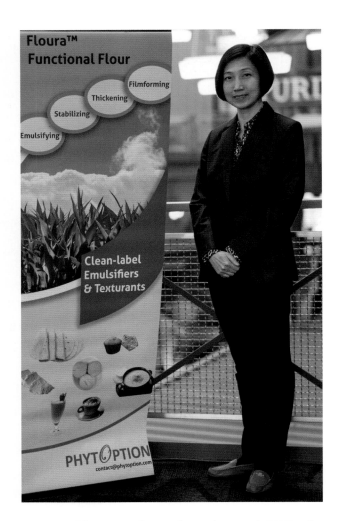

Joanne Zhang is the cofounder and CEO of Phytoption LLC. Photo courtesy of Joanne Zhang.

172

Zhang at the Indiana Conference for Women. Photo courtesy of Joanne Zhang.

After earning her bachelor's and master's degrees in food science from Jiangnan University in Wuxi, Zhang moved to the United States to pursue an MBA at Pennsylvania State University. For two decades she worked at large food corporations, including as marketing manager at Danisco (DuPont) and as finance director at Praxair. After gaining experience in the food industry in the fields of marketing, supply chain logistics, finance, and business unit management, she felt ready to start her own business.

In 2008 Zhang cofounded Phytopian LLC as a spin-off from Purdue University. The company develops natural and sustainable ingredients for food, beverages, and personal care products such as cosmetics. As of June 2020 there were five employees and the company had raised some $1.3 million in Small Business Innovation Research grants, including from the National Science Foundation. Their next focus is on selling the ingredients through food manufacturers and also specialized distributors in the food and personal care industries.

Although leading a foodtech start-up goes well beyond working 9 to 5, Zhang maintains a vast number of outside-the-job interests in food and culture. "I love visiting museums, trying new restaurants, and traveling," she says.

Q: What led you or inspired you to launch your company?

A: I am a food scientist by training. Working in corporates over the years, I have built management experience and expertise in operating businesses. After my daughter was born, I was constantly struggling to find natural foods or beverages without synthetic chemicals, especially being a scientist and knowing the origins of ingredients and their potential impact on human body. A seed was planted at that time, and I decided to do something to change the situation. When the opportunity arrived, I launched Phytoption.

Q: What led you to the agtech sector?

A: I fell in love with the food and agriculture business in college. Over the years, I continued to invest time into my curiosity surrounding this field. I attend food industry trade shows almost every year to keep myself up to date on the newest and best innovations.

Q: What problem or problems are your innovation solving and how?

A: Consumers love natural and healthy food and beverages. However, manufacturers of processed foods cannot make everything all natural due to the fact that they need to have an extended shelf life, contrary to homemade goods. Functional ingredients are required (natural ingredients such as what you would find in your own pantry are not functional enough to achieve that goal). We are changing this situation by using natural ingredients with strong functionality such as emulsifying and texturizing. They are natural and clean without any chemical treatment, gluten, or GMO.

Q: What inspires you to continue what you do?

A: We really see the market's needs for cleaner, healthier, and more simple ingredients. We are inspired by numerous consumer feedback demanding better ingredients in their food and personal care products. We feel obligated to carry through our mission—to make everyday products we consume healthier.

NANCY SCHELLHORN

Cofounder and CEO of RapidAIM Pty Ltd., Brisbane, Australia

Nancy Schellhorn was born and raised in the Midwest in the United States and grew up in Missouri and Texas. A city girl with a love of the outdoors, Nancy became interested in studying entomology when, while studying for her master's degree, she realized that insects compete with humans for food and that the ways in which insects are managed cause harm to the humans and the environment. Prior to her MS, she earned her bachelor's degree in animal science from the University of Missouri-Columbia. She also has earned a master's degree in ecology from the University of Missouri-St. Louis and a PhD in entomology from the University of Minnesota.

She landed in Narrabri, Australia, in 1999 working with Commonwealth Scientific and Industrial Research Organisation (CSIRO), Australia's national science agency. She was enticed by CSIRO's sound reputation, the adventure of Australia, and her then Australian boyfriend (and now husband), whom she met during her time at the University of Minnesota.

Schellhorn spent thirty years in agriculture, including time in landscape-scale pest management and at least ten of years in agtech. She launched her company, RapidAIM Pty Ltd., in September

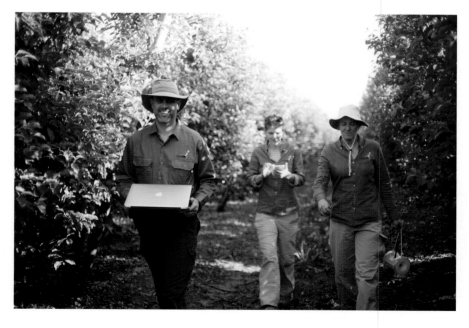

Nancy Schellhorn with fellow RapidAIM cofounders Darren Moore and Laura Jones. Photo courtesy of RapidAIM.

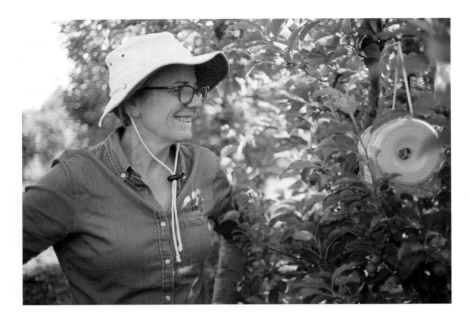

Schellhorn in the field during one of the company's first deployment of its sensors. Photo courtesy of Nancy Schellhorn.

2018 and since then she and her cofounders have raised US$950,000. As of 2020 RapidAIM has five staff.

The company's main product is the RapidAIM digital pest surveillance and management service. Customers subscribe to RapidAIM and receive real-time insect pest alerts across their orchards, farms, and other properties. The target customers are farmers, regional grower groups, and local and state governments across Australia.

Q: What led you or inspired you to launch your company?

A: Three things inspired me to launch it. Firstly, it's very exciting and fulfilling to achieve impact with our science. Secondly, our innovation solves a problem—we take the guesswork out of pest management—benefiting farmers and sustainable food production. And thirdly, my cofounders are amazing, so the ability to build a company together with tech that we created to solve a problem is a once-in-a-lifetime opportunity.

Q: What led you to the agtech sector?

A: The need to solve a basic problem about knowing where and when insect pests show up in orchards, farms, and regions. Combining biology and engineering is our magic and solution to the problem. We take the guesswork out of pest management.

Q: What is next for your company over the next year?

A: We want to start to scale our business both in terms of geography and adding additional revenue layers for key pest groups. Most crops have two or three key pests that cause most loss and cost. Our modular system can address many pest groups. Our first application of our hardware-enabled software is fruit fly, and our RapidFLY product. However, we'll soon be adding other pest groups to our product line, RapidMOTH and RapidBUG.

Q: What inspires you to continue what you do?

A: I know that we have the best technology to solve a global problem. We need to get RapidAIM in the hands of growers, farm managers, and biosecurity officers. RapidAIM will reduce cost and loss and help us meet our mission, which is to reduce and eliminate the dependency on chemical insecticide.

SARAH NOLET

Founder and CEO of AgThentic, New South Wales, Australia

Sarah Nolet's journey into the agtech space feels at times like kismet. The native Californian grew up in Silicon Valley where her parents worked in the semiconductor industry as chemical engineers. Although agriculture doesn't run in the family, her father purchased a hobby farm when she was twelve, and she ended up spending considerable time there as a teenager and into her twenties.

Sarah Nolet is the founder and CEO of AgThentic, an advisory and consulting firm, whose clients range from corporate agribusinesses to research organizations to investors to government. Photo courtesy of Sarah Nolet.

A gap year at an organic farm in South America was a turning point in Nolet's career. Photo courtesy of Sarah Nolet.

Nolet followed her parents' path and planned a career in engineering. She majored in computer science and human factors engineering at Tufts University in Massachusetts. After graduation she worked at a software company doing defense contracting for three years. As a systems engineer she traveled to military installations to build technologies to make intelligence analysis more efficient.

"I loved the technology aspects but wanted to get back into the private sector," says Nolet. She took a vacation, in part to give herself time to reflect, but the vacation fast extended into what she describes as a "gap year." She spent a year in South America, mostly living on an organic horticulture farm in Argentina.

That trip was a turning point in her career as she connected "the dots between the technologies we had been using in the defense industry, such as remote sensing, and the challenges in agriculture, such as weed detection," Nolet says.

In a separate interview she explained, "I absolutely loved it there. It made me reconsider what I wanted to do career-wise. I was pretty hooked on agriculture and decided I wanted to work in ag and get involved in helping develop technologies that could solve problems for farmers."

After a year living on farms in South America, she returned to the US, intent on a career in applying her technology skills to agriculture. She returned to school to earn a master's degree in system design and management from the Massachusetts Institute of Technology in 2016.

That same year, Nolet moved to Australia for her boyfriend David Shephard's job and also launched AgThentic, a consulting and advising firm for investors and entrepreneurs in the food and farming spaces.

"I literally didn't know anyone in Australia when I came," she says. "The next few years I was climbing a steep learning curve in building a business, and researching, and learning all I could about Australia's agtech space."

She observed that Australia is similar to California in farming systems and water challenges, but it is different in several key ways: it is highly export oriented and is the second least subsidized agriculture industry in the world. Additionally, it includes several unique production systems, such as the northern pastoral and mixed farming systems.

In less than five years Nolet founded three agtech companies, all in Australia. Following advisory firm AgThentic, she cofounded Tenacious Ventures in 2018, a venture capital fund that invests in agtech start-ups. Tenacious Ventures achieved first close at US$20 million in early 2020. She also cofounded Farmers2Founders, which runs innovation programs for farmers keen on tech adoption, or who want to build agtech or value-adding businesses.

On the personal front, Nolet loves playing sports (she was a competitive soccer player growing up and in college) and is now a competitive beach volleyball player, competing nationally in Australia.

"I live in Australia, there's lots of opportunity to get outside," she says, laughing.

Q: How did you get into agtech?

A: I grew up in Silicon Valley and my parents worked in the semiconductor industry. My dad bought a farm when I was around twelve. It was really a hobby farm, and we focused mostly on environmental restoration projects, but it did expose me to agriculture a bit. I mostly grew up playing sports, and so eating healthy food was important, as was the environment. But I really thought of these things more as a hobby than a career, so I initially studied technology and worked in the tech space.

Q: Why did you launch AgThentic?

A: I launched AgThentic because I realized there was a gap—so much technology was being built and pushed into the industry rather than pulled in. There was too much focus on the tech, and not enough focus on the users. I saw this as a huge problem because of all of the momentum agtech was experiencing, with new perspectives and new talent, but too many solutions were missing the mark. And at the same time, agriculture is so big and important—so this was really a shame. That is what led me to create AgThentic, helping [to] bring the best practices of new venture creation and innovation to agriculture while also making sure they could be adapted for agriculture so they could really have an impact.

Q: What inspires you to continue what you do?

A: What inspires me are the farmers and entrepreneurs that I am so privileged to work with. They are both really at the leading edge of what is truly going to change the world, and they are incredibly humble, resilient, hardworking, and inspirational people whom I just genuinely love to hang out with and be around every day.

SHEFALI MEHTA

Founder and CEO of Open Rivers Consulting Associates, Washington, DC

"Technology is the reason I exist," says Shefali Mehta, a first-generation South Asian and the founder of Open Rivers, an agtech consultant firm.

"My father immigrated to the US in 1970 to pursue a career in computer science and engineering. The US was one of the few places with such opportunities at that time." Once the family settled in Connecticut, her father and her mother (who studied physics) found work on an assembly line. Her mother stopped working to focus on taking care of the household during the recession in the 1970s.

Many of Mehta's family worked in science or technology—an uncle and a cousin were rocket scientists, "which drove my passion from an early age." Both sides of her family placed a strong value on the role of "science and mathematics in our lives," she says.

Shefali Mehta enjoyed connecting with growers such as Minnesota farmer Brian Ryberg in her previous position as executive director of the Soil Health Partnership. Photo courtesy of Shefali Mehta.

It is of little surprise that Mehta, who was born and grew up in Bethel, Connecticut, picked a path that emphasized higher education and STEM. She earned a bachelor's in economics from New York University, a master's in economics from the University of Cambridge, and a master's in statistics at the University of Minnesota.

But agriculture also captured her interest from the time she was a girl. During a trip to India during her college days, it struck her that smallholders had an "absolute lack of transfer of knowledge and technology."

"I realized that we needed to do more, so I jumped into the fray," Mehta says of her decision to commit herself to a career in the agriculture sector. She earned a PhD in agricultural and applied economics from the University of Minnesota. Her dissertation focused on managing invasive species in US forests, lion hunting, and conservation.

Her career following her education included full-time roles at McKinsey and Syngenta. Mehta has also served as a mentor to young women interested in STEM by serving as a board member of the School of Statistics at the University of Minnesota and the Techstars Farm to Fork Accelerator.

In 2017 Mehta launched Open Rivers Consulting Associates out of her base in northern Virginia to provide technology solutions to clients in food, farming, and the environment. The company's current clients include the Foundation for Food and Agricultural Research (FFAR) and the US Farmers and Ranchers Alliance (USFRA). It also does pro bono work for organizations whose missions focus on soil health and climate resilience.

The following year she took a detour to lead and relaunch Ceres Wave, an agtech company utilizing plasma to increase crop productivity, which brought plasma, used throughout other industries, into agriculture in a novel and impactful way.

She also led the Soil Health Partnership, a farmer-led nonprofit organized under the National Corn Growers Association that identifies the impact of conservation practices on soil health through data and science.

In 2019 Mehta relaunched Open Rivers (what she refers to as Open Rivers 2.0), later adding a small team of research assistants. As of June 2020, the company has been working on data, digitization, and technology in agricultural and food, including with FFAR.

We have also been focused on creating and implementing a novel data landscape scan for a client Agriculture Climate Partnership, established by FFAR and USFRA, alongside many other organizations.

She enjoys being a connector and using her network to help make a positive difference.

"What I do is who I am and vice versa—everything reinforces and feeds into this broader mission so even hobbies such as diving, kayaking, running, photography, cooking, and gardening are actually what ties me and connects me to the communities, people, and the world that I endeavor to support and strengthen," she says. The business reflects Mehta's do-good intentions by offering an Open Rivers Scholarship that supports women going into STEM or environmental studies—there

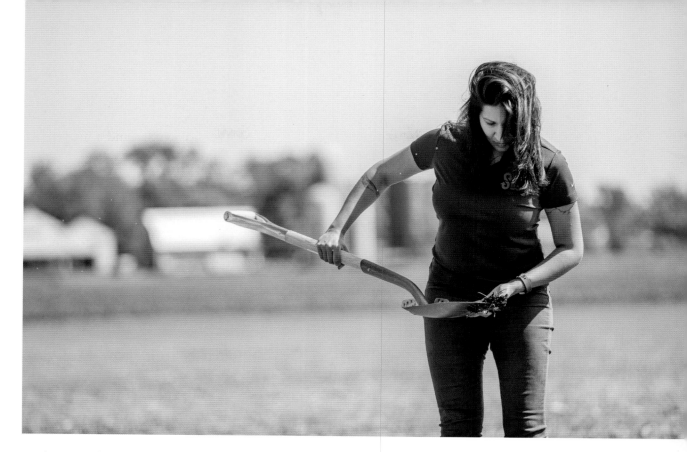

Mehta in a field on Ryberg Farms in Buffalo Lake, Minnesota, viewing the impact of soil practices on soil texture. Photo courtesy of Shefali Mehta.

have been eight scholarships disseminated to date.

And technology, she feels strongly, has always been a pivotal part of solving significant problems in the food and farming systems.

"Technology is critical to ensure that all farmers are able to economically, environmentally, and sustainably grow food. Technology is also critical to addressing how we feed a growing population while restoring and nurturing a healthy environment," she says.

When not working, Mehta loves attending concerts ranging from classical music to rock ("I used to work at an indie rock label in New York City"), scuba diving, hiking, boating, and spending time with friends and family.

Epilogue

The march toward innovation in agriculture continues at a fast clip, in part fueled by COVID-19. As the pandemic descended globally and spread fast, I sheltered in place in my apartment in the Hudson Valley, New York. The importance of freshly grown food and its availability became even more important. I felt blessed to be based in a region that has always been known for its family farms and open access to fresh and local produce.

In the wake of COVID, there seemed to be an explosion of interest in fresh and local food among community members and neighbors and city folks who had relocated to the countryside and in understanding the origins of the food they consumed. In the spring the CSAs (community-supported agriculture) were oversubscribed, the farmers markets were abuzz, and the community garden had a lengthy waitlist. This had already been happening pre-COVID with a shift toward organically grown food and efforts to connect community with local agriculture. An example is Woodland, in California's Yolo County, which launched The Food Front, a platform that connects local restaurants and farmers with consumers through resources and events. Efforts such as these are a slant of sun for a sector that had traditionally taken a backseat to other sectors. Agriculture hasn't been sexy, but things were starting to shift.

After connecting the dots between consumers, farmers, technologists, and innovators, I concluded that storytelling is essential if those connections are to be made obvious to the general public.

On this journey of searching for women founders in agtech, I'd developed a curiosity as a consumer and in time a deep appreciation for the growers and production staff who are devoted to feeding the communities they live in. I had an opportunity to do so first as a journalist and then in communicating farming to the general public.

In spending time on farms I saw firsthand the challenges of growing food in vast quantities and fighting weeds and pests, especially through organic practices. In tending to my own little garden, I found myself torn between killing pests to save my potatoes and understanding that the insects that invaded the vegetables were also seeking sustenance. In observing the pests closely, I also noticed their beauty. The Colorado potato beetle has a striking lime and black-striped coat that makes it almost look dapper.

I marveled at the amount of time and investment it took to transform soil from subpar to healthy. It takes years, along with a tremendous amount of science, research, knowledge, and money. After a season on a farm and overseeing my own garden, I commented to a good friend, "I think I am too old for this. This is hard labor." Indeed, seeding, planting, mulching, weeding, and harvesting were tedious jobs that required a tremendous amount of physical stamina. I felt like a wimp.

Although there was no single formula for growing beautiful fruits and vegetables, farming requires precision in light, temperature, or any number of other factors. Mother Nature could easily throw a wrench in one's plans at any given moment. Although there was no single aha moment for me, I increasingly became aware how much agriculture and innovation were clearly intertwined and how one depends on the other.

The word in Chinese for *crisis* consists of the characters for danger and opportunity. The women innovators who have shared their stories with me saw an opportunity and took advantage of it. Some moved across the state or even across the country to pursue their vision.

A visit to Knoll Farm–Brave Little Farm in Waitsfield, Vermont, in the summer 2020. Photo courtesy of Amy Wu.

Upon reflection, though, this is not surprising. Many were already pioneers in their own right—the first from their family to receive their education in the US, the first to attain multiple higher degrees, the first to achieve key leadership roles at the companies or organizations where they had worked. It takes chutzpah to forge one's own path and to break past barriers, be they cultural or financial. Women and minorities have historically been underrepresented in both ag and tech industries. Now, this new generation of start-ups led by minority women is seeking to solve agriculture's problems with tech innovation.

Despite their diverse backgrounds and unique stories, the women shared numerous commonalities, including, but not limited to, the following:

- A passion for changing the world for the better through their technology
- A passion to solve big problems with impact
- A desire to give back and create a community of like-minded people
- A desire to get away from corporate politics to a more merit-based environment and to be able to better execute their vision
- A passion for creating something new and innovative that others have not done or been able to do

Simply put, they have taken the ball and run with it. The question now was whether investors and growers would welcome their innovations with an open mind.

What Now? What Next?

Since 2016 *From Farms to Incubators* has grown from a series of stories to an initiative that combines storytelling and advocacy.

The handful of women innovators I originally found now number in the dozens, and every week someone forwards me another email and asks, "Have you interviewed this woman?" I can't help but feel a thrill when I receive these questions.

Back in 2016 I'd attend ag conferences and observe skepticism among many growers about agtech. They'd question the effectiveness of everything from sensors to robots with the same air someone might have toward snake oil. Since then growers have not only become more open to the idea of new innovation but have begun to seek innovation out. By 2020 the fast spread of the COVID-19 pandemic and the resulting furious pace of change in the world made clear that adopting and adapting to innovation would become necessary if farms were to not only survive but thrive.

Although certain sectors of the economy faltered during COVID-19, investment in agtech seemed to pick up steam, not just in the US but globally. There were some high-profile investments: in May 2020 Apeel Sciences, a Santa Barbara–based company that's developing technology to extend the freshness of produce, raised $250 million in new financing led by GIC and Upfront Ventures, Viking Global Investors, and celebrities Oprah Winfrey and Katy Perry. Tech giant Microsoft has stepped up its investment in creating data-driven farming for precision agriculture, with the goal of helping smallholder farms. One of the results is Azure FarmBeats, a cloud platform that aggregates farming data from different sources, including satellites, sensors, and drones. In Salinas, California, the Western Growers Center for Innovation and Technology announced the Government of Canada as its first international partner.

Agtech accelerators and incubators were popping up in places that one might not traditionally associate agriculture with. In Kentucky, for example, agtech is the state's top economic development priority for the goal of creating more jobs, and the state has partnered with some twenty organizations to develop what it envisions to be "America's AgTech Capital in Appalachia."[1] And Crusonia on the Delta (formerly Davos on the Delta), an annual gathering of investors, industry experts, and growers in Memphis, Tennessee, has become one of the most popular agtech events in the US.

Agtech developments in nascent stages were winning grants, and many of those leading the developments were women. In July 2020 Nadia Shakoor, a senior researcher at the Donald Danforth Plant Science Center, was awarded a $1.4 million grant from the National Institute for Food and Agriculture and the National Science Foundation to develop the FieldDock, an agtech drone base station.[2]

"People still need to eat, of course, and want affordability, sustainability, and transparency in food production," says Pam Marrone of Marrone Bio Innovations. "Chemical pesticides continue to be removed from the market by regulators around the world, providing opportunities for companies. I continue to hear about and be contacted by new agtech and agbio women entrepreneurs. There are still many big problems to solve and there is still money being invested in start-ups despite COVID-19."

The women whom I first met in 2016 felt that things were moving in the right direction, for both the sector and female leaders.

"I am definitely seeing more women in the agriculture space in general. I think it's only a matter of time before more women start getting into technology as well," says Jessica Gonzalez of Happy Organics. The sector would eventually mature to a point where diversity and inclusion would extend into serving underserved communities.

During the spring of 2020 Gonzalez launched a mobile app and online directory of agricultural businesses led by BIPOC (Black, Indigenous, and People of Color). "I do know that we need more agtech that serves underserved communities. Most agtech focuses on the bottom line for

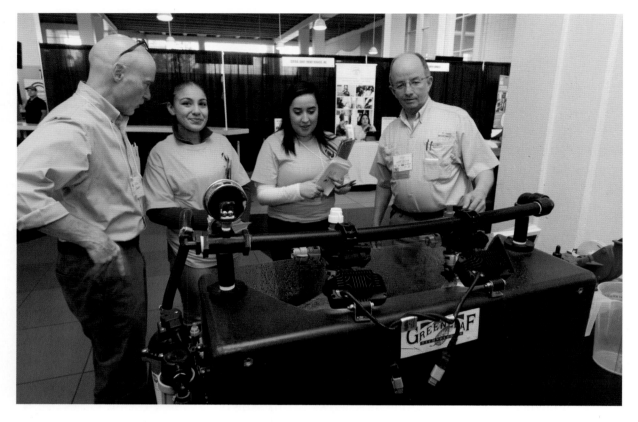

Innovations being showcased at the Salinas Valley AgTech Summit. Photo courtesy of Richard Green.

commercial agriculture. There is a huge need for food justice in low-income communities that are typically BIPOC. I see myself gravitating toward filling this need in the tools that I am building," she says.

In the meantime, challenges continue to mount for farmers. Growers face constant pressure to feed a global population that is expected to reach nine billion people by 2050. The labor shortage extends as fieldworkers age out. It is questionable how much arable land will remain for future generations to farm. It is even more questionable who will farm in the future as the cost of business rises, and many young people seek professions that are more lucrative and less labor intensive.

In writing this book, I became aware of an unspoken hope that youth will read these stories and understand that opportunities in agriculture extend far beyond tractors and overalls. It is artificial intelligence, blockchain, drones, sensors, robotics, and automation. The future is bright for youth who seek opportunities in agriculture and technology, as well as in the corresponding areas of agbio and foodtech.

Uplifting New Voices

The journey in seeking the stories of women founders in agtech has taught me that it is critical to continue to seek stories that are underreported and undertold. By sharing and highlighting those stories we uplift new voices and we fill in the gaps of what will someday be history. It is important to understand history if we are to connect with the present and consider the future—who owns the land and who is farming the food we eat.

"I think projects like this one are great for amplifying the stories of women in agriculture. Combined with the success of female founders in the space, I think it is a vertical that is encouraging for female founders to get into," says Ellie Symes, CEO of The Bee Corp.

Christine Su, founder of PastureMap, says women leaders in agriculture and technology have come a long way, but there is a way go to. "For BIPOC women entrepreneurs, there's never been a better time to bring our whole selves to the work. We are living in a food and ag tech ecosystem that was imagined by and primarily for white men—look where it's gotten us. It's scary to keep pointing out where this system doesn't work for all of us, when the space is still dominated by them. But we must be brave because we owe it to the next generation of women."

Finally, this story is an immigrant story. We often view the history of agriculture in the US through a mainstream lens, ignoring the contributions of people who have been marginalized but who are nonetheless part of the country's social fabric. It is important to see the history of contributions of everyone. And for me this started with asking a question and continuing to asking the same question repeatedly: "Do you know of any women entrepreneurs in agtech, especially women of color?"

As the late associate justice Ruth Bader Ginsburg stated, "Real change, enduring change, happens one step at a time."

With this last chapter, one part of the journey comes to an end, but the stories continue to emerge and evolve. I hope that others will continue on the next stage of the journey with me.

Finally, this book represents the efforts of the pioneers in the field of agriculture and technology up to 2020. And of course innovations and advances in these fields will continue. For those who are interested in furthering their study of this growing industry, please visit www.farmstoincubators.com.

—Amy Wu, August 2020

A Final Word from the Author

Dear Reader,

From Farms to Incubators continues its mission to tell the stories of women leaders in agtech. If you know of a woman leader or founder in agtech, or agbio, please email me details at amy@farmstoincubators.com.

If you'd like to be added to our newsletter mailing list or are interested in learning more about the *From Farms to Incubators* documentary, please, sign up at www.farmstoincubators.com, where you can also find the latest news on the women profiled in this book and updates related to the intiative.

Thank you again for supporting the initiative.

Notes

1. Brian Sparks, "Why Appalachia Might Become the Next Greenhouse Tech Hub," *Greenhouse Grower*, June 28, 2020, https://www.greenhousegrower.com/production/why-appalachia-might-become-the-next-greenhouse-tech-hub.

2. Chris Albrecht, "FieldDock Project Gets $1.4M Grant for AgTech Drone Base Station," The Spoon, July 17, 2020, https://thespoontech/fielddock-project-gets-1-4m-grant-for-agtech-drone-base-station.

Carved lima bean Farms to Incubators logo.
Courtesy of Sergey Jivetin.

The Road Ahead

Afterword by Danielle Nierenberg

I get asked a lot about what I think is the most pressing issue in the food system—from Food Tank members, from the press, from young people and students, and from other advocates and researchers. And I could give a whole long list of problems, ranging from COVID-19 and climate change to soil degradation and from the absence of nutrient density in crops to obesity. But the one issue that stands out for me is inequality and the lack of attention, research, and funding for women farmers, inventors, scientists, and entrepreneurs, particularly those who are Black, Indigenous, People of Color, and Latinx. Women are consistently overlooked for their contributions in agriculture, especially when it comes to innovation and technology.

Globally, women make up 43% of the agricultural labor force. In some countries in sub-Saharan Africa, for example, women make up 70 to 80% of all farmers. But they often are denied access to the same resources as their male counterparts, including land, credit, financial resources, education, and extension services, and—maybe most critical of all—respect.

The same is true in the agricultural sciences and technology. In sub-Saharan Africa, only one in four agricultural researchers is female. Fewer than one-fourth of executives in the food industry are women, and fewer still are women of color. According to recent research from Karen Karp & Partners, in the US in 2017,

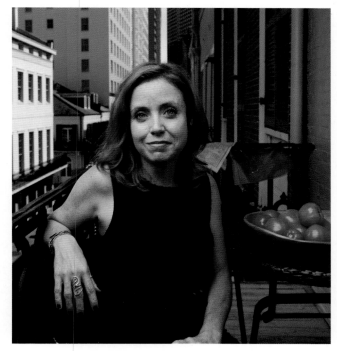

Photo provided by Danielle Nierenberg.

female founders of agri-enterprises received just two percent of venture capital investment, less than $2 billion of the $85 billion total investment across all sectors of agriculture.

One thing that is clear, however, is that we ignore women in the food system at our own peril. According to the U.N. Food and Agriculture Organization, if women had the same access to resources as men, they could raise their current yields by 20 to 30%—this would lift as many as 150 million people out of hunger.

But despite the challenges women in the food system face, they are also generating enormous success. The agtech sector, in particular, offers women opportunities to create healthier, more nutritious, and more environmentally and economically sustainable food systems across the globe.

While many of the stories in *From Farms to Incubators* focus on California, the book also gives readers a snapshot of what is happening globally in fields such as artificial intelligence, drone technology, water and soil sensors, data analysis, weather tracking, and marketing and research.

The stories show how women are not just pioneering food and technology innovations but providing inspiration to their communities. Generations Y and Z are more interested in food and agriculture than ever before, and the intersection of food and tech is helping make agriculture cool and a career path that youth see opportunity in as well as a chance to change the world.

Young women—and men— who envision creative ways to transform agriculture while also acknowledging the benefits of traditional knowledge and indigenous practices. The combination of high- and low-tech innovations has never been more important than today as the next generation of farmers and innovators work to address soil degradation, erratic weather, water scarcity, and other challenges impacting food production.

There's no doubt that technological innovations will continue to develop, changing how we grow, harvest, process, distribute, market, sell, eat, share, recover, and dispose of food.

I think the next decade offers enormous challenges for women in all parts of agtech and agbio. Inequality continues to pervade how women are educated, mentored, and how they gain opportunities. But the struggles we've seen over the past year for achieving equity and social justice are not only inspiring and powerful, but leading to real change for women and so many other groups that have been overlooked. And the pandemic has lifted the veil on what has been hidden in the food system for so long—cruel injustice and inefficient supply chains. Women tech experts will be an important part of fixing and eliminating those problems.

Danielle Nierenberg is the cofounder and president of Food Tank, a global convener, research organization, and non-biased creator of original research impacting the food system. Nierenberg has an M.S. in Agriculture, Food, and Environment from the Tufts University Friedman School of Nutrition Science and Policy, and spent two years volunteering for the Peace Corps in the Dominican Republic.

Acknowledgments

The African proverb "It takes a village to raise a child" is certainly applicable to *From Farms to Incubators*, which started as a journalism series that spawned a documentary that led to this book and to the initiative it is today.

This book is the result of support from a community of friends, families, mentors, acquaintances, colleagues, and fellow journalists who understood why I pursued the story from the start and why I kept at it. It is a given that I am grateful to the book's contributors, talented photographers, artists, and editors and to the sponsors/supporters who helped make this book happen. At the same time it would not have been possible without the support from those who contributed their talent, time, ideas, support, inspiration, and encouragement to keep me going.

Thank you to the following people and organizations that supported the initiative from the start: the City of Salinas, the City of Gonzales, the Western Growers Center for Innovation and Technology, DigitalNEST, David Leighton of Women in Technology International, Javier Zamora of JSM Organics, Pam Marrone of Marrone Bio Innovations, Karen Caplan of Frieda's Specialty Produce, Jonathan Eppers and Marc Staudenbaur of VYBES and then Inspira Studios, Green Thumb Organics, the Urban Arts Collective, Happy Organics, Downtown Book and Sound, and Knoll Farm.

The following organizations and entities played unique roles in providing support in various ways: AgStart, California State University, Monterey Bay, *Coastal Grower* magazine, Ecological Farming Association, Food Tank, Hartnell College, and the VINE.

I am especially grateful to the International Center for Journalists and International Media Women's Foundation's Howard G. Buffett Fund, which gave the initiative a solid foundation.

Many people made this book happen, and I have room to highlight only a few. To my father, Dr. Joseph M. Wu, who taught me to dream big and follow my passions. He was one of the earliest supporters of From Farms to Incubators, both the initiative and the book. At his birthday celebration on August 1, 2018, he thanked me for celebrating with him and encouraged me to "believe and be a visionary." Two years later he reminded me of the following:

Believe in your interest and passion
Believe in your goals and aspirations
Believe in the power of the Almighty
Keep the gift of self-respect and connectivity

He is a pioneer in his own right and has created so many opportunities and better lives for his family, and for the many graduate and medical students who benefited from being taught or advised by him.

Thank you to my stepmother (TC) for encouraging me to complete the book no matter what and to my sister, Mary, for making life colorful and sweet. To my cat, Yul, who makes the best writing companion. Thank you to Uncle William and Auntie Jane for your love and mentorship and for reminding me to keep telling stories. To Grandma, Auntie Yvonne, and Uncle Sam, who are always with me in spirit.

Thank you to the many friends on both coasts who cheered me on along the journey, whether by attending the screenings or encouraging me to complete the book—Susana Vazquez (who also donated her talent in translating some of the content into Spanish), Katharine Ball, Pallavi Guha, Merrill Perlman, Pastor Ben Sobels at Cypress Church in Monterey, Christine Bottaro, Gil and Aprille Lucero, Albert and Ginger Fong, Jim Helm, Steven Berkowsky, Joey Givens, Douglas Stewart (and his beloved late dog, Willie), TW Leung, Pete Wevurski, Jessica Trites Rolle, Marie Cole, my colleagues at the Refresh Policy Platform, and the team at the Ecological Farming Association—especially Andy Fisher, Deborah Yashar, and Gabi Salazar. Gratitude to colleagues at the Hudson Valley Farm Hub. Gratitude to Kara Goldin of HINT Water for her support.

Thank you to journalists Mary Duan and Sara Rubin at *Monterey County Weekly*, the journalists at the *Salinas Californian* and the *Salinas Valley Tribune*, and the editorial teams at *Techonomy* and *Worth*, who continue to give me a platform to share the stories. To the organizations who gave Farms to Incubators a further voice, including Women in Agribusiness, Thought For Food, the Salinas Valley Chamber of Commerce, and Paul Farmer and the Silicon Valley Forum. A special thank you to Ed Edwards and the team at KSBW, which covers the Central Coast, for their excellent coverage.

A tremendous thank you to Kent Sorsky, the publisher of Linden Publishing, who took a gamble on a new topic and was always there to answer my questions. Within the *Farms to Incubators* team I am especially grateful to Beret Erway for proofreading each and every word and for her witty edits, which served as both inspiration and learning. On the visual front, a special acknowledgment to Dexter Farm, who not only contributed beautiful photos but kept the project in check with his professionalism and a shared vision for this project.

A special thank you to the following people and organizations for generously allowing us to use their photos for the book: John Cerney, Richard Green Photography, Benjamin Hsu, Chava

Oropesa, Eugenia Renteria, Melanie Smelcher, Salinas Valley AgTech Summit, SVG Partners, Doug Westcott, Western Growers Center for Innovation and Technology, and World AgriTech Summit.

Thank you for Renee Edelman for your friendship and for believing in the project and helping me bring it to *Techonomy*. A special thank you to Dennis Donohue, who answered every call, and to Ray Corpuz Jr., who supported the initiative from the start.

I am also grateful to have worked with Michele Speich, Natalia Luna, Ernesto Altamarino, and the team at the National Steinbeck Center, who immediately supported the corresponding exhibition once they heard about the book. Tremendous gratitude to the artists who contributed their art to the exhibition and book, including J.C. Gonzalez (for your shared passion), Dexter Farm, Daphnee Parachini, Le Vuong, and Sergey Jivetin.

Thank you to the filmmakers who were a part of the exhibition, including Garrett Stern, John Picklap, Jackie Mow, Laura Pacheco, and Frieda's Specialty Produce's team, who shared *Fear No Fruit: The Frieda Caplan Documentary*. Thanks to Gaibi Dhiman, who helped on the back end of the Farms to Incubators website, whether it be posting stories or launching new pages for the exhibition and book.

Gratitude to Brad Barbeau, Shyam Kamath, Mary Jo Zenk, and the team from Startup Monterey Bay Tech Meetup for all of your support. Thank you to Aaron Magenheim, CEO and founder of AgTech Insight, who consistently connected me with women in agtech. Thank you to Mrs. Hiller (Lana Hiller) and Donna Garr from my high school days for encouraging me to keep writing. Thank you to my fellow swimmers who kept me sane here in California and New York.

A special tribute to Mayor Joe Gunter of Salinas, who suddenly passed away on June 29, 2020. I enjoyed working with him when I reported for the *Californian*. He always answered my calls and spent considerable time helping me get acquainted with a city that he was clearly devoted to. I still fondly remember him taking half a day at least to give me a tour of the city, including the farms and neighborhoods where many farmworkers lived, when I first arrived. And long after I no longer worked at the newspaper and lived in Salinas, he continued to encourage my work in agtech.

Upon reflecting on this project, we also celebrate the women in agriculture who were pioneers in their own right: Frieda Caplan, Kay Hiatt, and Dolores Huerta, to name a few.

Now there is a new generation of trailblazers. Most of all I am indebted to the women who shared their stories. They are pioneers in their own right and have been essential in writing a new chapter in history. They motivated to me to keep writing and keep asking questions when things looked rough. I hope they will not only influence more women to share their stories but inspire the next generation to follow in their footsteps and forge their own path.

My father told me that a book in many ways is a personal passion, and once it's complete one can move on. With the ending here I don't move on but instead continue to forge ahead. There are many more stories to share.

Contributors to
From Farms to Incubators

The following people graciously lent their time and talent to this book. I am immensely appreciative of what they contributed.

William Au (photographer) is trained in engineering and has been an amateur photographer for over twenty years. Born in Hong Kong and raised in Macau, he came to San Francisco to attend college and later finished his degree in the California State University, Sacramento. He has been a civil engineer for more than forty years, and his engineering career gives him the opportunity to bring his camera while working on projects across California. His favorite subject to photograph is the beloved garden of his wife, Jane.

Beret Erway (copy editor) is an experienced copy editor, covering finance (The Deal.com), insurance (*Insurance Advocate*), and medical and pharmaceuticals (ME Data guides, *Physicians' Desk Reference*). On the side, she has also done freelance work in art, biography, and science fiction.

Dexter J. Farm (photographer) is an award–winning self-taught photographer. Dexter's focus includes macro (extreme close-up photography) landscapes, sea life (whales), and local community events in Monterey County, California. When he is not taking pictures, Dexter is out reviewing the local wineries.

JC Gonzalez (artist) is a visual interdisciplinary and community-based artist who uses the power of art as a portal for community voice. JC is the creative founder of The Urban Arts Collaborative (UAC), based in Salinas, California, a multi-disciplinary, socially conscious arts organization that weaves interconnected issues of critical importance for the community, such as creative expression, food justice, the environment, racial equity, and youth leadership.

Sergey Jivetin (artist) is a New York-based artist who focuses on presenting miniature elements in unexpected settings to examine humanity's convoluted relationship with nature. He is the recipient of numerous accolades, including fellowships from the Louis Comfort Tiffany Foundation, and his work is in the permanent collections of many public and private entities, such as the Smithsonian Institution and the Metropolitan Museum of Art. His ecology-themed artworks are online at www. sergeyjivetin.com, and the Furrow Seed Engraving Project can be found at www.seedengraving.com.

Eugenia Renteria (filmmaker) is a director, cinematographer, and editor based in Watsonville, California. Most of her work is centered around her life experiences as an immigrant woman. Born and raised in a small town in Zacatecas, Mexico, she moved to California when she was twelve, which is when her interest in filmmaking started. Shortly after graduating from California State University, Monterey Bay with a bachelor's degree in cinematic arts and technology, she cofounded her production company, Inspira Studios.

Susana Vázquez (Spanish translator) is currently working as a social services and health coordinator in the Salinas Valley region. She is experienced in child care, journalism, and social services and is fluent in English and Spanish. She has a passion for sustainability and upcycling.

Trav Williams (photographer) is a photographer who spends most of his time taking photos of and documenting the agrarian world. He works with nonprofits, businesses, and individuals to create long-term records of their work and lives. Between shooting and editing, he attempts to keep the weeds down on his own farm on the Oregon's North Coast.

Bill Winters (photographer) is an Emmy Award–winning director of photography with over twenty years of experience in the film and television business. He is known for creating the look of *Comedians in Cars Getting Coffee* for Jerry Seinfeld, frequent collaborations with Oscar nominated director Joe Berlinger, and his work with Oprah Winfrey's *Oscar Special* and several Netflix documentaries.

Deborah Yashar (copy editor) is the marketing and communications director and host of the Ecological Farming Association (EcoFarm), the oldest and largest organic farming conference in California. She earned her bachelor's degree in environmental studies from University of California, Santa Cruz, where she also earned a Merrill College Achievement Award for acquisition of Spanish and Portuguese and for her service to social justice efforts in Latin America.

Glossary

The following definitions were curated from a variety of sources, which are cited at the end of the glossary.

agbio: Companies that focus on using biotechnology to develop or improve biopesticides, plant growth efficiency, crop agriculture, and food technologies.

agtech: An industry at the intersection of agriculture and technology. Agriculture-related products and/or services that contain or are enabled by patented technology.

aquaponics: A system of aquaculture in which the waste produced by farmed fish or other aquatic animals supplies nutrients for plants grown hydroponically, which in turn purify the water.

angel investor: A person who invests in a new or small-business venture, providing capital for start-up or expansion.

artificial intelligence (AI): The simulation of human intelligence processes by machines, especially computer systems.

BIPOC: Black, Indigenous, and People of Color.

biotechnology: The manipulation (such as genetic engineering) of living organisms or their components to produce useful commercial products, including bacterial strains or pharmaceuticals.

blockchain: A digital record of transactions where individual records, called blocks, are linked together in single list, called a chain. Blockchains are used for recording transactions made with cryptocurrencies, including Bitcoin.

business accelerator: A program that provides advice, guidance, and various forms of support for businesses in the start-up phase. A business accelerator compresses the timescale for starting up by operating in the boot camp format. Companies that use business accelerators are typically start-ups that have moved beyond the earliest stages of getting established.

business incubator: A program that gives developing companies access to mentorship, investors, and other support that can help them become stable, self-sufficient businesses.

Central Coast, California: The geographical region that runs 350 miles along the coastline from Ventura just north of Los Angeles and Monterey County to Santa Clara south of San Francisco.

Central Valley, California: The geographical region that runs for more than 400 miles down the middle of the state and includes the cities of Bakersfield, Fresno, and Stockton.

cleantech: Technology that places an emphasis on environmentally friendly products, services, or practices.

conventional farming: Conventional farming that uses seeds that have been genetically altered by a variety of traditional breeding methods, excluding biotechnology, and are not certified as organic.

cover crops: Fast-growing crops, including rye, buckwheat, cowpea, and vetch, that are planted to prevent soil erosion, increase nutrients in the soil, and provide organic matter.

community-supported agriculture (CSA): A food production and distribution system that directly connects farmers and consumers. In short, people buy "shares" of a farm's harvest in advance and then receive a portion of the crops as they're harvested.

crop yield: A measurement of the amount of agricultural production harvested (the yield of a crop) per unit of land area.

DNA sequencing: Determining the order of the four chemical building blocks (called bases) that make up the DNA (deoxyribonucleic acid) molecule. The sequence tells scientists the kind of genetic information that is carried in a particular DNA segment.

fintech: A term used to describe financial technology, an industry encompassing any kind of technology in financial services, from businesses to consumers. Fintech describes any company that provides financial services through software or other technology and includes anything from mobile payment apps to cryptocurrency.

foodtech: The intersection between food and technology; the application of technology to improve agriculture and food production, the supply chain, and the distribution channel.

gene editing: The ability to make highly specific changes in the DNA sequence of a living organism, essentially customizing its genetic makeup. Gene editing is performed using enzymes, particularly nucleases that have been engineered to target a specific DNA sequence, where they introduce cuts into the DNA strands, enabling the removal of existing DNA and the insertion of replacement DNA.

genomics: The study of genes and their functions and related techniques.

grass-fed-and-finished: A diet of nothing but grass and forage for the animal's entire life, with no supplemental grain feed. If an animal was just grass fed, then it could be fed supplemental grain feed and/or finished on a fully grain-based diet. Finishing refers to the animal's final weight before it is slaughtered.

harvest: The season for gathering in agricultural crops or the act or process of gathering in a crop.

internet of things (IoT): An extension of the internet and other network connections to different sensors and devices—or "things"—affording even simple objects, such as lightbulbs, locks, and vents, a higher degree of computing and analytical capabilities.

machine learning: The application of artificial intelligence (AI) that provides systems the ability to automatically learn and improve from experience without being explicitly programmed. Machine learning focuses on the development of computer programs that can access data and use it learn for themselves.

nitrogen, phosphorus, and potassium: The three gases that are the primary nutrients in commercial fertilizers. Also known as NPK.

no-till farming: The practice of planting crops without tilling the soil. Also known as zero tillage.

The Natural Resources Conservation Service (NRCS): A service that provides incentives to farmers, ranchers, and forest landowners wanting to put wetlands, agricultural land, grasslands, and forests under long-term easements.

organic farming: An agricultural system that uses ecologically based pest controls and biological fertilizers derived largely from animal and plant wastes and nitrogen-fixing cover crops.

organic foods: USDA-certified organic foods grown and processed according to federal guidelines addressing, among many factors, soil quality, animal raising practices, pest and weed control, and use of additives. Organic producers rely on natural substances and physical, mechanical, or biologically based farming methods to the fullest extent possible.

pitchdeck: A presentation that entrepreneurs put together when seeking a round of financing from investors. On average, pitch decks have no more than nineteen slides.

pollinator: Anything that helps carry pollen from the male part of the flower to the female part of the same or another flower. The movement of pollen must occur for the plant to become fertilized and produce fruits, seeds, and young plants. Some plants are self-pollinating, while others may be fertilized by pollen carried by wind or water. Other flowers are pollinated by insects and animals, such as bees, wasps, moths, butterflies, birds, flies, and small mammals, including bats.

precision agriculture: Using new technologies to increase crop yields and profitability while lowering the levels of traditional inputs needed to grow crops (land, water, fertilizer, herbicides, and insecticides).

regenerative agriculture: Farming and grazing practices that, among other benefits, reverse climate change by rebuilding soil organic matter and restoring degraded soil biodiversity, resulting in both carbon drawdown and improving the water cycle.

Salinas Valley, California: One of the most productive agricultural regions in California. Nicknamed the "Salad Bowl of the World." The Valley includes the cities of Gonzales, Greenfield, King City, Salinas, and Soledad.

seed saving: The practice of saving various types of reproductive material and seeds from plants such as flowers, herbs, grains, vegetables, and tubers for future use. The traditional way that gardens and farms were maintained centuries ago.

soil sampling: Estimating the capacity of the soil to provide adequate amounts of the necessary nutrients to meet the needs of the crop (or crops) to be grown.

sustainable agriculture: A farming system [or method] that over the long term enhances the environmental quality and the resource base on which agriculture depends; provides for basic human food and fiber needs; is economically viable; and enhances the quality of life for farmers and society as a whole.

tillage, reduced tillage, and minimum tillage: The mechanical manipulation of the soil for the purpose of crop production, affecting significantly the soil characteristics such as soil water conservation, soil temperature, infiltration, and evapotranspiration processes.

United States Department of Agriculture (USDA): The federal agency that proposes programs and implements policies and regulations related to American farming, forestry, ranching, food quality, and nutrition. President Abraham Lincoln founded the USDA in 1862, when about half of all Americans lived on farms.

venture capital: A form of private equity and a type of financing that investors provide to start-up companies and small businesses that are believed to have long-term growth potential. Venture capital generally comes from well-off investors, investment banks, and any other financial institutions.

vertical: A business term that describes a specific industry or market that focuses on a particular niche.

Sources

The American Society of Agronomy, The Cary Institute, Investopedia, maximumyield.com, National Human Genome Research Institute, National Parks Service, Regeneration International, The Spruce Eats, sustainableamerica.org, United States Department of Agriculture, World Health Organization.

www.bioflorida.com, www.lexico.com, www.northeastsare.org, www.merriam-webster.com, www.thestreet.com, www.ussc.edu.au, www.visitcalifornia.com.

Supporters of
From Farms to Incubators

A special thank you goes out to the following supporters who helped make this book happen with their sponsorship, talent, and shared passion for uplifting the voices of women innovators in agtech.

City of Salinas
City of Gonzales
Frieda's Specialty Produce
Pam Marrone
VYBES
Renee S. Edelman
Women in Technology International

JSM Organics
AgStart and the AgTech Innovation Alliance
DigitalNest
Downtown Book and Sound
Ecological Farming Association
Green Thumb Organics
Inspira Studios
Knoll Farm–Brave Little Farm
National Steinbeck Center
Paicines Ranch
Urban Arts Collaborative
Western Growers Center for Innovation and Technology (WGCIT)
Dr. Joseph M. Wu and Tze-chen Hsieh
Mary H. Wu
Steven Berkowsky

INDEX

About the Author

Amy Wu is an award-winning writer for the women's ag and agtech movement. She is the creator and chief content director of From Farms to Incubators, a multimedia platform that uses documentary, video, photography, and the written word to tell the stories of women leaders and innovators in agtech.

From Farms to Incubators has a mission of highlighting women in food, farming, and farmtech, especially women of color. From Farms to Incubators includes the documentary film and book *From Farms to Incubators*, which spotlight women leaders in ag and agtech. The documentary and stories have been screened and presented at SXSW and Techonomy. The initiative was awarded grants from the International Center for Journalists and the International Women's Media Foundation's Howard G. Buffett Fund for Women Journalists. In 2019 Wu was named one of Food Tank's "15 Inspiring Women Leading the AgTech Sector" and in 2020 appeared *Worth* magazine's "Groundbreakers 2020 of 50 Women Changing the World" list. Wu has also received the Women in Agribusiness Demeter Award of Excellence.

Prior to starting From Farms to Incubators, Wu spent over two decades as an investigative reporter at media outlets, including the *USA Today* Network, where she reported on agriculture and agtech for the *Salinas Californian*. She's also worked at *Time* magazine and the *Deal* and contributed to the *New York Times*, the *Huffington Post*, *Forbes Women*, and the *Wall Street Journal*.

Wu earned her bachelor's degree in history from New York University and her master's degree in journalism from Columbia University. An avid open-water swimmer, she completed the 28.6-mile Manhattan Island Marathon swim in 2010. When not writing and swimming, she enjoys spending time with family, friends, and her cat, Yul Brynner.